PROCEEDINGS

OF THE

ELMIRA ACADEMY OF SCIENCES.

Elmira, N. Y.
PUBLISHED BY THE ACADEMY.
1892.

Reprinted by New York History Review
Elmira, New York
ᖷ 2014 ᖷ

Elmira Academy of Sciences, 1892
Reprinted by New York History Review

ISBN: 978-1-312-26326-0

Printed in the United States of America.

Table of Contents

THE ELMIRA ACADEMY OF SCIENCES had its origin in 1858, in the labor and liberality of Prof. C. S. Farrar, of the Female College in this city, together with a few public-spirited citizens.

Practical astronomy was their first idea. The grounds for a building were donated by Hon. E. P. Brooks. About $2000 were subscribed for the building an observatory; telescopes and other apparatus were purchased; considerable debt was incurred, which in a few years, was cleared off.

In the west wing was placed a good transit telescope. In the center stands a sideral clock and a museum of minerals and curiosities. In the east wing are placed an electric chronograph and a small library. The dome above contains a fine, refracting telescope, equatorially mounted; its length is 113 inches, with a clear aperture of 8 ½inches; it has seven Huyghenian eye-pieces, commanding powers of from 55 to 880, and has the usual circles, reading microscopes, and clock-work movement.

The academicians (numbering about fifty gentlemen) hold business meetings at stated periods, and hold scientific meetings as occasion demands, at the call of the president. At these meetings certain standing committees report and discuss scientific matters in their departments, and generally two or more members present papers on special subjects of investigation; often the evening is spent in inspecting specimens of geology or natural history, or in examining some new instrument of philosophical research.

The society usually reports its meetings in the current local news of the day. A small and valuable monograph, on "The Birds of Southern New York," by one of its officers, is its only publication as yet. A collection of its scientific papers and proceedings will probably be published ere long.

From the beginning, having no endowment fund nor income to support an able astronomer who might give his whole time to the work of discovery, nor having any convenient hall for meetings, the society has aimed chiefly to promote the diffusion of scientific knowledge, and the culture of a taste and aptitude for scientific pursuits rather than original discovery. There has indeed been the purpose and preparation for adding a scientific hall to the observatory, where lectures, experiments, and discussions on the natural sciences and education might be held practically few to all; but the city is yet young, and members of scientific taste and sufficient wealth to bring this about are too few. A considerable amount of useful and interesting work has already been done by the society.

From *The History of Tioga, Chemung, Tompkins, and Schuyler Counties New York,* 1879.

A SKETCH OF THE ORIGIN

OF THE

ELMIRA ACADEMY OF SCIENCES

And its Succeeding History.

Early in the year 1859, a few gentlemen of a scientific turn of mind, citizens of Elmira, became interested in the formation of a society for the mutual benefit of its members in scientific research—more particularly the astronomical branch. Subscriptions were solicited for the purchase of an Astronomical Telescope of moderate size. This effort fortunately met with more than the expected interest and encouragement, and it was decided to purchase a Refractor of eight and one-half inches clear aperature, of Henry Fitz, Esq., of New York.

This instrument was brought to Elmira, June 2nd, 1859, and was first mounted upon a parallactic ladder in the garden of Prof. C. S. Farrar, of Elmira College. Here it remained until an observatory building being indispensably necessary for its satisfactory use, more subscriptions were solicited and obtained. This money was used in the expenses of the observatory, while payment for the telescope was secured to Mr. Fitz, by a chattel mortgage running for ten years. The ground required for the erection of the observatory was generously leased by Hon. E. P.

Brooks, being a triangular park at the intersection of Park Place and College Avenue, near by the Elmira Female College. The lease was granted for one hundred years, with no other restriction than that the land should be used explicitly for the purpose specified. The building commenced in August, of 1859, was finished, ready for occupancy, April 1st, 1860.

The credit for the management of all this preliminary work is due entirely to Prof. Charles S. Farrar. The success of the enterprise is a monument of his energy, skill and self-sacrifice. The property and contributions amounting to some two thousand dollars were vested in him. At this juncture he proposed to those who had so ably assisted his efforts to equip the observatory, the formation of a society, appropriate to the spirit of the movement thus far, to whom he would deed the property, in trust for future work in scientific education.

A meeting was held at the observatory library June 8th, 1860, at which the following gentlemen were present: Prof. C. S. Farrar, Dr. Wm. H. Gregg, Dr. I. F. Hart, J. R. Ward, M. H. Arnot and Dr. T. H. Squire. A committee consisting of Prof. Farrar, Dr. Gregg and Mr. Ward was appointed to draft a constitution, and submit a name for the proposed society. At further meetings the organization was completed by the adoption of the name of "The Elmira Academy of Sciences," and a constitution and by-laws for its government. The first officers were:

President—PROF. CHARLES S. FARRAR.
Vice-President—DAVID MURDOCH, D. D.
Secretary—FRANK H. ATKINSON.
Treasurer—SAMUEL R. VAN CAMPEN.
Curator—WILLIAM H. GREGG, M. D.,
And a Board of Trustees, ten in number.

This society held regular meetings during the following months until May, 1861, when the formal proposition was made

for the incorporation of the society, and its assuming the owner-
ship of the property. This certificate of incorporation was
granted June 27th, 1861. A list of the incorporators appears in
the certificate published in this pamphlet. It is unfortunate
that a full list of the membership of the Academy was not
obtained at this period of the organization. Many gentlemen
took part in the meetings, contributing largely to their interest
who are not known as actual members. Elmira has been fortu-
nate in having in her citizens, gentlemen who have been eminently qualified to contribute much useful instruction to such gather-
ings as have attended the meetings of the Academy of Sciences.
The plan of work adopted by the organizers has been in the
main that of success.

In addition to the fine refractor, purchased at the time of organization, the Academy soon added a fine chronometer, costing
three hundred dollars, and a magnificent transit instrument and
circles, of one thousand dollars more, these fully equipping the
observatory for all astronomical work.

In March, 1876, the observatory building was entered at night
and the transit was stolen. It was fortunately recovered in New
York where it had been left by the thief, without any apparent
damage to its mechanism.

In January, 1884, the Academy discussed at length the pro-
priety of making a deed of its real estate and personal property
to the Elmira College, reserving to its members forever the use of
the astronomical observatory upon its land, for its meetings, and
scientific pursuits, and reserving forever the library, museum of
specimens, and the geological collection, with the right to retain
them where they now are. This was finally done by vote, January 14th, 1884. This generous act has furnished the Elmira
College with a fully equipped astronomical observatory for the
instruction of the students without any other expense that the

maintainance of the property in repair. The meetings of the Academy are held in the large lecture room of the observatory and the students of the College are admitted to them, joining frequently in the exercises.

At a meeting held January 7th, 1886, the Elmira Microscopical Society was united with the Academy, becoming a section of the same, to be known as "The UpDeGraff Microscopical Club" of the Elmira Academy of Sciences. They have ever since been active workers with the society, assisting materially in the public meeting.

The purpose of the founders of the society as before stated has been continued through a period of over thirty years. If at some periods in its history, the general public has not been so enthusiasic in attendance upon its meetings, the interest of individual members has not slackened nor has its treasury ever been lacking in ability to meet its expenses. The publication of its doings has not extended beyond the circulation of the newspapers of the city, who have furnished interesting reports of its meetings.

The following will be of interest as showing the work of the Academy and the range of topics discussed at its meetings :

The first reported scientific meeting was held August 13th, 1860, Rev. Thomas K. Beecher, chairman of the committee upon "Physical Science", at the regular meeting in September, read an essay upon "The Course to be pursued by the members to ensure the best work from the Academy".

Notably, essays or papers have been presented on scientific topics by the following : Prof. Farrar, Geological explanation of the Crust of the Earth ; Rev. David Murdoch, on the changes in the shore-line of the Gulf of Mexico ; Prof. Ford, Geometry of plants and law of philotaxis—The Chronograph ; Mr. Augustus McConnell, Geology of Chemung Co.; Mr. Francis Collingwood, The Time-ball as dropped by electric machinery—The Brooklyn Bridge—Has the Earth ever rotated on a different axis from its present one ; Dr. Wm. M. Gregg, On Vegetable respiration—Prof. I. M. Wellington, How does Nature sow her forests ? or why does hardwood spring up after a pine forest is cut away ? Rev. Dr. A. W. Cowles, Antiquity of Human remains ; Prof. Rogers, Rating of Elmira observatory, and longitude of Elmira, which is long. 52.6s. East of Washington ; Prof. Farrar, Asteroids—Sun and Sun-spots—Tycho Brache ; Rev. Thos. K. Beecher, Gunnery—"Horology" or how I found the true time out in the woods ; Mr. McConnell, Geology of Canada ; Prof. Wellington, Climates, as investigated and recorded by the Ancients, beginning with Ptolemy ; Mr. Francis Collingwood, Musical chords—Laws of Rest and Motion—Necessity of Forest culture in the United States ; Prof. James E. Latimer, Insanity; Dr. I. F. Hart, Vaccination and its prevention of Smallpox ; Prof. Bentley, Oceans ; Dr. Wm. M. Gregg, Habits and peculiraities of Crow-Black birds, (molo thrus peccoris). Prof. L. C. Foster, Steam valves and cut-offs—Unfolding of repetends ; Dr. UpDeGraff, Embalming of birds ; Prof. Ford, Anaestheties —Analysis of Colors—Silver ores and methods of reducing them —The Philosophy of colors—Sun-spots—The Modern application of Electricity and its use as a motor on railways (Feb. 12, 1892)—Mind in animals ; Dr. H. D. Wey, The Microscope in evidence ; Dr. Adele A. Gleason, History of Errors in Microscopic work ; Prof. Charles A. Collin, Land Tenure ; Mr. Francis Hall, Alaska—Yellowstone Park ; Mr. W. N. Eastabrook, Bicycling in Europe ; Dr. T. F. Lucy, Flora of Chemung County; Mr. I. B. Coleman, Electric welding ; Mr. Frederic Hall, Mind Cure and kindred delusion ; Mr. R. R. Moss, Weregild, or the pecuniary value of the social atom ; Dr. F. W. Ross, Role of micro-organism in disease.

In conclusion, the officers of the Academy acknowledge that many difficulties attend the supervision of a society which has so high sounding a title as this. They propose to make the study of the several departments of science a benefit to the members. To the older and much larger fellow-societies they appeal for a place in the ranks of knowledge-gatherers represented by them. In the few printed researches published by us we endeavor to fur-nish information worthy of attention, and evidencing that we are on the line of scientific investigation which shall benefit all comprised in the terms of our chartered title.

R. A. HALL,
Secretary.

THE ANURONA GLEASONII--An Annelid of the Nais Tribe,--By T. S. Up De Graff, M. D., F. R. M. S. Elmira, N. Y.

This curious water worm has been found at different times both in winter and in summer in a large wooden tank in my conservatory, used for keeping alive *Mesobranchi* and other reptiles. Its favorite haunt seems to be upon the under side of a board sunken in the bottom of the tank and covered with the various forms of the rotifera. To the unaided eye it appears as a long, slender, white worm, about one-half of an inch in length. Examined under the microscope, under a half-inch objective and a one inch eye-piece it presents the following characteristics : Body pellucid, annulate and numbering thirty segments that are shorter but larger in transverse diameter as they approach the ventral portion. The cephalic segment is lanceolate and fringed with delicate vibratory cilia. Rudimentary, black, eye-spot discernible near the center of this segment. At the articulation with the next segment is located, on the under side, a triagular mouth with contractile labia from between which, in the act of feeding, is protruded a fandibuliform proboscis which grasps the food particles, consisting of algae and desmids, withdrawing them within the esophagus that leads to an elongated stomach. The segments are supplied with chitinous setae, arranged in four rows—two dorsal and two ventral—its groups of three, save the six ventral segments of the cephalic portion, where the setae become uncinated and are arranged in groups of five on each side of the central surface of the segments. These setae and hooklets are partly contractiee by means of fan-shaped muscles that have their attachments within the segments. They are used as aids to locomotion, and are employed much as are the legs of worm-like animals. The vent is situated in the center of the terminal segment, and is surrounded with six ciliated tentaculiform processes or lobes equidistant from each other. These lobes are club-shaped, nodose, and fringed with rather coarse, rythmically vibrating cilia that produce strong whirling currents in the water, directing floating particles toward the ventral orifice. These lobes are contractile, capable of being wholly withdrawn within the body. The perivisceral cavity is occupied by a straw-colored fluid containing pinkish globules of varying sizes that move freely about through the vermicular action of the intestinal canal and of the body of the animal. The intestinal tube, for a distance extending to about the sixteenth segment from the anal aperture, is lined with very delicate cilia, from the movement of which a current of water is directed inward, carrying with it minute monads and other food particles. Whether this ciliated structure is designed for purposes of respiration or an aid to nutrition, or for both purposes combined could not be determined in the short period for observation afforded by the specimens for examination.

This annelid is perhaps the handsomest and most interesting of its class, presenting a very beautiful appearance under a one-inch objective, when its six plume-like tentacles are fully ex tended and the fringed margins in rapid vibration. I regret my inability to have observed more fully its digestive and circulatory apparatus, that I might have added a description of them also. I think that some new provisions for these functions will be therein revealed. No description answering to that of this curious annelid is to be found in any of the literature to my command. In Dr. McDonald's "Water Analysis" is figured one somewhat similar, having four lobes. There being no descriptive text, but simply the brief announcement that it is "conformable with the Proto of Oken." I am unable to determine how many features it may have in common with the one herewith presented.

A New Anuræa—This rotifer differs entirely from any hitherto described, more particularly in its possession of a dorsal horn or spine. The lorica is in form of an oblong square, with the two transverse angles, posteriorily, acutely truncated, on the dorsal aspect. From the four acute angles thus formed project, with a slight inward curve, four spines of nearly equal length and attenuated at the free extremities. In the median line is situated a horn, curved slightly backward, and of a length corresponding to one-half the length of the lorica. Immediately in front of this dorsal horn is placed a bright, scarlet eye. The ventral side of the lorica is obtusely oval, terminating posteriorily in a long spinous horn, straight and extending to an acute point. There are in all six spines; four at the four posterior angles of the dorsal surface, one on the dorsum in the cervical region and one on the ventral extremity of the lorica. The anterior margin of the shell is spineless. The brow extends slightly beyond the border of the lorica, is doubly arched on both surfaces, from between which the animal protrudes two short, trancated lobes, from within the lorica, and near the lateral brims, upon which the rotatory organs are situated. Otherwise, the animal possesses the usual characteristics of its conquers, in the possession of and location of biliary glands, ovaries, three-toothed jaws, oesophageal head, alimentary canal, and stomach. The lorica is arched dorsally and rough. Ventral surface, flat and smooth. The creature revolves almost constantly on its anteris-posterior diameter. Length of lorica, independent of the spine, 1-145th of an inch. It is found in stagnant ponds, and is most abundant in July, although I have taken specimens from under the ice in mid-winter. I regard it as a unique specimen of the Brachionæa family, by reason of its dorsal spine, five posterior spines, and without anterior spines. Believing it to be new and hitherto undescribed, I have named it the Anuroca Gleasonli, in honor of the President of the Elmira Microscopical Society.

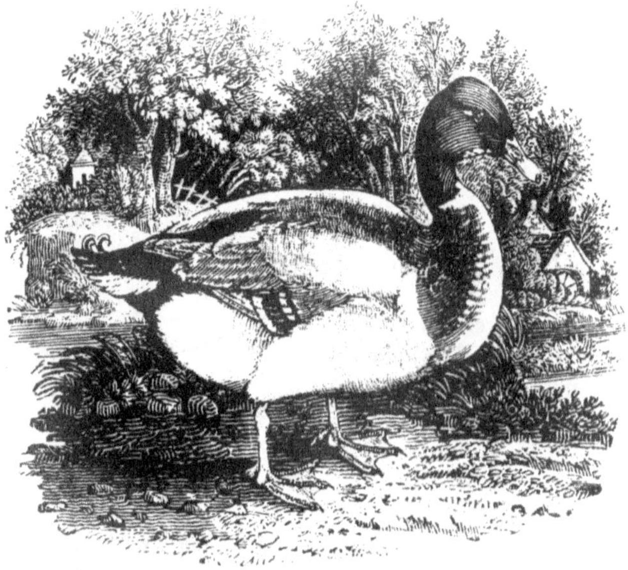

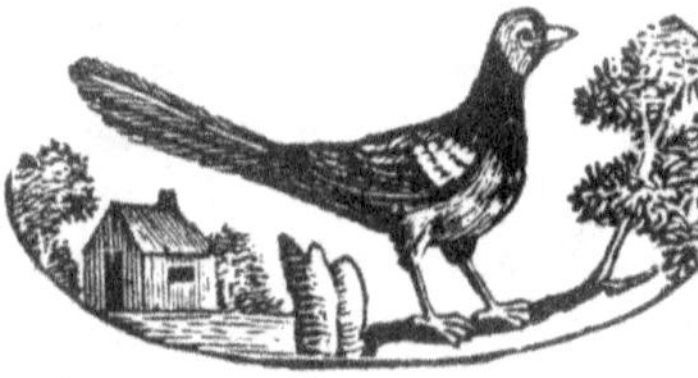

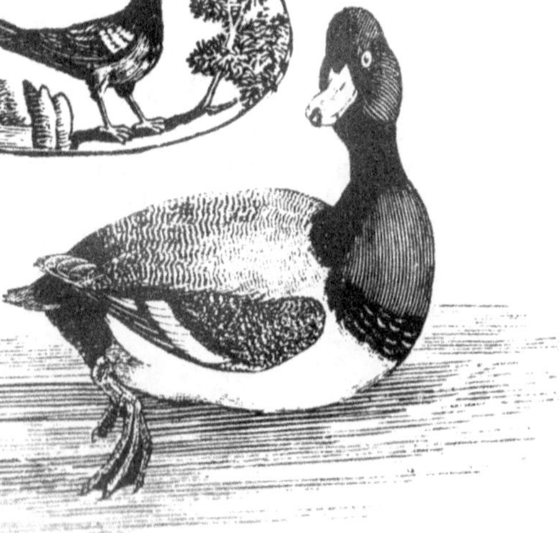

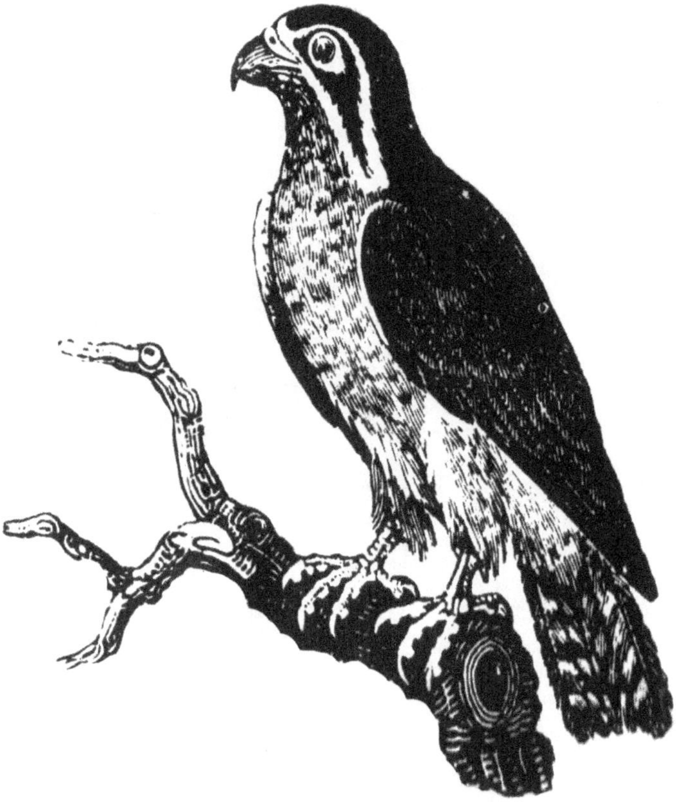

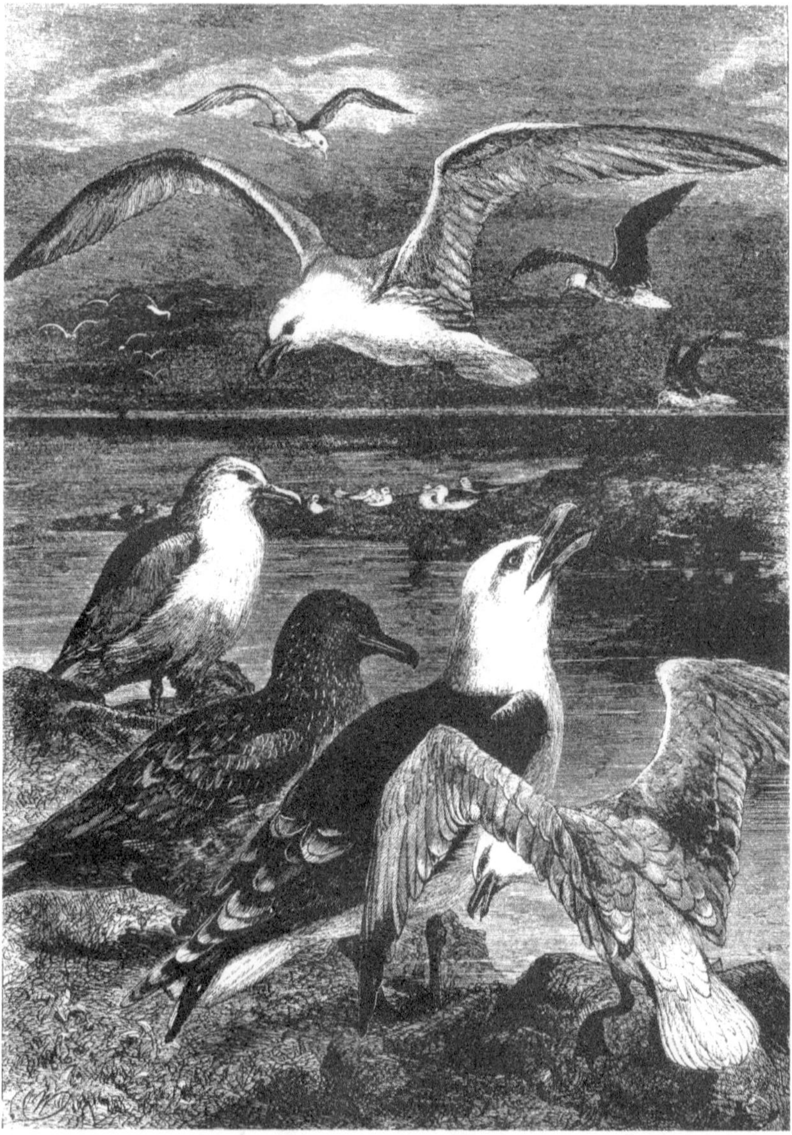

CATALOGUE of the Birds of Chemung County.--By WM. H. GREGG, M. D.

The following catalogue of birds, incidental to Chemung county, comprises the results of several years of close observation. Some species are included of which I have no personal knowledge of their occurrence in this locality, but as they are found in localities not very remote, I do not hesitate to include them, feeling that they will yet be met with within our limits.

Our county is physically well adapted for harboring the numerous species which are included in the list; the extensive marshes north of us, as well as the river and its estuaries, afford ample and well fitted localities for the wading and aquatic species, while our extensive valley and mountain ranges offer enticing retreats for the numerous land birds which are shown by the following list to visit us.

The number of species of birds found in America, north of Mexico, amounts to about 720. Of this number about 25 are extra limital, giving a total of North American birds at about 740. In this county, so far as our observation extends, we have about 196 species, and I have no doubt that the number will be swelled to 200 or over, when the facts are known. Our list is meager, when compared with more favored localities on the sea border, where the number is increased by the addition of aquatic species, that never proceed inland as far as this point; still our number is large when compared with the whole number incidental to the State.

The central location of our county, between the extreme north and south, marks it as a favorable point for ornithological study. One of the most interesting facts connected with the study of birds, is their migration. It is well known that they love to follow the river courses, or pursue their migratory journey along the base of mountain ranges and in the valleys of our continent. For this reason many species which are natives of America, west of the Rocky Mountains, follow the Pacific slope to the vicinity of Mackenzie's river, crossing the mountains over an elevated plateau of that region, and spread out over the British possessions, from the Arctic Circle to the Atlantic district of Canada, while stragglers from these migratory bands are known to wander to this and neighboring States.

It has also been observed that species which are summer residents of the territory east of the mountains pursue a course along the Atlantic coast, east of the Appalachian range, or through the valley of the Mississippi, and are not liable to journey beyond, unless driven out of their course by uncontrollable causes. Eastern species, which are known to winter in Mexico, crowd into the Southern States at the approach of winter, and reach their winter quarters in Central America by way of the Gulf. Many do not venture a journey over the sea, but continue their course by land, while other species pass on to the West Indies as a stopping place, and finally renew their journey, to winter in tropical America. There are some species which do not seem to be governed by any particular migratory law, but are spread far and wide, over the whole extent of the continent.

In order to study more closely the distribution of species incidental to this locality, it becomes necessary to limit our observation, that we may not be misled on account of extent of territory. It is by keeping within certain limits that we shall sooner arrive at a knowledge of the correct geographical distribution of species.

To suit our present convenience, we will describe a circle, the circumference of which shall reach about one hundred miles from (this city) the center. Beyond this line species occur which never reach this locality, and by thus limiting our territory we shall be able to arrive at a more correct knowledge of the extent and distribution of our fauna. This imaginary line would spread out at the north to within a few miles of Lake Ontario, east to Sullivan county, south to near Sunbury, Pa., and west almost to the shores of Lake Erie.

I have not been able, for want of time, to map out the precise line of migration, or the extent of distribution of species which would become subjects of our observation.

It is unnecessary for me to enter upon any speculation or attempt to explain why certain species should shun certain sections

of country which to all appearances is as well fitted for their nidi-
fication, growth and development as locations only a few miles
distant. It has been attributed to the effect of certain conditions
of the atmosphere, in regard to heat and moisture. Temperature
must certainly be one of the controlling causes. It is only by
studying closely the habits and distribution of birds that we shall
be able to arrive at any correct explanation of the causes of their
migration.

The number of constant resident species—summer and winter
residents—and those which pause for a few days only, on their
passage north and south, together with other facts connected
with the ornithology of our county, will be considered in a future
communication.

1. *Falco columbarius*, (Linn.)—Pigeon-Hawk. A species of common
occurrence in this locality, and is found throughout temperate North
America. I am not aware it is a constant resident, although I have noticed it as late as November and also early in March.

2. *Falco sparverius*, (Linn.)—Sparrow Hawk. This is a common species,
occurring throughout the continent of America. It affects open woods
bordering on cultivated fields, and is the smallest reptorial bird inhabiting
the United States.

3. *Astur articapillus*, (Wilson.)—The Goshawk. One of the rarest of
the hawks met with in this locality. It is a northern species, visiting us
during fall and winter.

4. *Accipter cooperi*, (Bonap.)—Cooper's Hawk. A common species, frequently met with in this locality.

5. *Accipter fuscus*, (Gmel.)—Sharp Shinned Hawk. This species inhabits America, from the Arctic regions to Mexico. It is often met with in
this section of the State.

6. *Buteo borealis*, (Gmel.)—Red Tailed Hawk. This species is common
in all Eastern North America. It ranges from the fur countries to the
West Indies.

7. *Buteo lineatus*, (Gmel.)—Red Shouldered Hawk. This and the pre-
ceding species are the most common of the hawk tribe that occur in this
locality. B. lineatus builds a nest of very rough construction, in the forked
branches of the oak, and usually lays three eggs. The young may be distinguished before they leave the nest by the red patch on the shoulder.

8. *Buteo pennsylvanicus*. (Wilson.)—Broad-winged Hawk. I have not
been able to obtain a species, although I have no doubt it will be found
here.

9. *Archibuto lagopus*, (Gmel.)—Rough Legged Hawk. This species is not of uncommon occurrence. It is usually met with in low, marshy situations, and is common to Europe and America.

10. *Archibuto sancti Johannas*, (Gmel.)—Black Hawk. A well known species inhabiting America, from Pennsylvania north. It is more frequent in the low valleys and along water courses.

11. *Circus hudsonicus*, (Linn.)—Marsh Hawk. A common species during summer.

12. *Aquilla canadensis*, (Linn.)—Golden Eagle. The Golden Eagle is not a common bird in this locality; several specimens, however, are known to have been obtained here. It occurs at all seasons of the year.

13. *Haliaetus leucocephalus*, (Linn.)—Bald Eagle. Specimens of this species are of constant occurrence during summer.

14. *Pandion carolinensis*, (Gmel.)—Osprey, Fish Hawk. Of late years this species has become quite common along our creeks and rivers. We should hear more of it were it not that it is often confounded with the above—H. leucocephalus.

15. *Bubo virginianus*, (Gmel.)—Great Horned Owl. This large and elegant species is well known, from the loud and dismal noise which it makes at all times of the day and night.

16.—*Scops asio*, (Savig.)—Screech Owl. This is a common and well known species.

17. *Otus wilsonianus*. (Lesson.)—Long Eared Owl. This is not a very common species in this locality.

18. *Brachyotus cassinii*, (Brewer.)—Short Eared Owl. This is a winter visitant, and not of unfrequent occurrence.

19. *Surnium nebulosum*, (Foster.—Barred Owl. This species is of common occurrence, and is frequently seen flying about during the day.

20. *Nyctale acadica*, (Gmel.)—Saw Wheet Owl. This small and familiar species is of common occurrence. It nests in the hollow of trees, some - times in the forked branches, preferring inaccessible swampy situations. Usually lays five eggs.

21. *Nyctea nivea*, (Daub.)—Snowy Owl. This large and handsome bird is not often met with in this locality. It is a northern species, only visiting us during severe winters.

22. *Coccygus americanus*, (Bonap.)—Yellow-billed Cockoo. A species of common occurrence. Inhabits orchards and open woods.

23. *Coccygus erythrophthalmus*, (Bon.)—Black-billed Cuckoo. I can dis-

cover but little difference in regard to the frequency of this and the above species. There seems to be but little difference. Both species are quite common.

24. *Picus villosus*, (Linn.)—Hairy Woodpecker. This familiar little species is a constant resident of this State.

25. *Picus pubescens*, (Linn.)—Downy Woodpecker. This is a common species, resident throughout the year.

26. *Sphyrapicus varius*, (Baird.)—Yellow-bellied Woodpecker. This is a remarkably scarce species in this locality. I know of but two specimens being taken here.

27. *Centurus carolinus*, (Bonap.)—Red-bellied Woodpecker. This handsome bird is not met with as often as formerly. The clearing up of the forests has decreased the number of the Woodpecker tribe with us.

28. *Melanerpes erythrocephalus*, (Swainson.)—Red-headed Woodpecker. This species is not as common as formerly.

29. *Colaptes auratus*, (Swains.)—High Holder. This species is very plentiful, and seems determined not to fall back as the forest disappears. It retains its hold in the orchard and groves.

30. *Trochilus colubris*, (Linn.)—Ruby-throated Humming Bird. This beautiful little species is met with from early spring until late in the fall. It is very abundant with us, and is the only species of its tribe that visits us.

31. *Chaetura pelasgia*, (Steph.)—Chimney Swallow. This is a well known and abundant species.

32. *Antrostomus vociferus*, (Bonap.)—The Whippoorwill. This bird, whose plaintive whistle is heard at all hours of the night, is an abundant and favorite species.

33. *Chordeiles popetue*, (Vieill.)—Night Hawk. This familiar species is very common, particularly during July and August.

34. *Ceryle Alcyon*, (Boie.)—King Fisher. This is the only species of the genus that inhabits the United States east of the Mississippi. It is common in this locality.

35. *Tyrannus carolinensis*, (Baird.)—King Bird. The King Bird or Bee-eater is a very common species and is one of the most pugnacious of our birds, making battle even with the falcons and other species much larger than itself.

36. *Myiarchus crinisus*, (Cabanis.)—Great-crested King Bird. A common species, not so frequently observed, however, as the above, on account of its more retiring habits.

37. *Sayornis fuscus*, (Baird.)—Phœba Bird. This common little fly-catcher

is abundant in this locality, building its nests under bridges, in old buildings. It delights in water courses on account of the abundance of its insect food.

38. *Contropus borealis*, (Baird.)—Olive-sided Fly-catcher. Cooper's King Bird. This is not a very common species with us. I have met with only two specimens during several years of bird collecting.

39. *Contropus virens*, (Linn.) - Wood Pewee. Common in this locality and is an abundant species throughout the Northern States.

40. *Empidonax traillii*, (Baird.)

41. *Empidonax minimus*, (Baird.)—Least Fly-catcher. This is a common species, and is one of the most active little birds that visit us.

42. *Empidonax acadicus*, (Baird). Small Green Fly Catcher. A common though not abundant species,

43. *Empidonax flaviventris*, (Baird).—All the species of Empidonax delight in damp, secluded situations.

44 *Turdus mustelinus*, (Gmel). Wood Robin. The sweet notes of this bird enliven our woods and groves from early spring until October frosts. It is a common species.

45. *Turdus pallasii*, (Cabanis). Hermit Thrush. This is a very common species in our open woods.

46. *Turdus fuscescens*, (Stephens). Wilson's Thrush. This species is resident during summer, retiring early in the fall. It prefers dry open woods, and keeps much on the ground.

47. *Turdus swainsoni*, (Cab). Olive backed Thrush. A common summer resident and abundant species.

48. *Turdus migratorius*, (Linn.) Robin. This species is a constant resident. In winter they retire to the deep recesses of the forests. Their food at this time consists of the larva of insects and the seeds and berries of the flowering plants with which our forests abound.

49. *Sialia sialis*, (Baird.) Blue Bird.—This species arrives early in March and remains until late in autumn. It is a common and abundant species.

50. *Regulus calendula*, (Licht.) Ruby Crowned Kinglet. There is considerable difference between the habits of tne Ruby and Golden crowned Kinglet. The former bird partakes in fact, more of the habits of the Fly catcher, by capturing its prey upon the wing, and living among the higher branches of tall trees, where it skips from branch to branch in chase of its insect food. The note of the Ruby crowned is quite loud and long continued, resembling somewhat that of the Warbling Greenlet (*V gilvus*) while

the note of the Golden crown is a low sharp twitter, resembling very much, the call of *Particapillus.* They are more shy of the approach of man than the Golden crowned.

51. *Regulus satrapa,* (Licht.) Golden crowned Kinglet. This species approaches nearer the habits of the Tits, spending most of its time on the middling and lower branches of the pines ; running up or down at pleasure upon the perendicular branches hunting out the hiding places of its favorite food in the crevices of the bark or amongst the mosses.

52. *Anthus ludovicianus,* (Licht.) Tit lark. Sky Lark. A winter visitant ; not of common occurrence.

53. *Miniotilta vara,* (Vieill.) Black and White Creeper. An abundant species, arriving during the last days of April.

54. *Parula americana,* (Bonap.) Blue yellow backed Warbler. A common species remaining during summer.

55. *Geothlypis trichas,* (Cab.) Maryland Yellow Throat. A common species, affecting low damp situations.

56. *Geothlypis philadelphia,* (Baird.)—Mourning Warbler. Male and female shot May 24th, 1870.

57. *Goethlypis agilis,* (Baird.) Connecticut Warbler. This species has not been noticed, although it should be found here.

58. *Oporornis formosas,* (Baird.) Kentucky Warbler. No specimens taken.

59. *Icteria viridis,* (Bonap.) Yellow Breasted Chat. This is not a common occurring species, although specimens are obtained every season.

60. *Helmitherus vermivorus,* (Bonap.)—Worm-eating Warbler. This species is not of common occurrence.

61. *Helminthopaga pinus,* (Baird.)—Blue-winged Yellow Warbler, Female, shot May 22d, 1867.

62. *Helminthopaga chrysoptera,* (Cab.)—Golden-winged Warbler. Rarely obtained. Specimen shot May 22d, 1867.

63. *Helminthophaga ruficapilla,* (Baird.)—Nashville Warbler. I have met with several specimens of this species. It is not rare with us.

64. *Helminthopaga peregrina,* (Cab.)—Tennessee Warbler. I have not been able to obtain a specimen of this rare species ; I know of but one specimen being taken here.

65. *Seiurus aurocapillus,* (Swains.)—Oven Bird. A common and abundant species.

66. *Seiurus noveboracensis,* (Nuttall.)

67. *Seiurus ludovicianus,* (Bonap.)—Long-billed Water Thrush.

68. *Dendroica virens,* (Baird.)—Black-throated Green Warbler. A common species, during summer, met with in damp oak and maple woods

69. *Dendroica canadensis,* (Baird.)—Black-throated Blue Warbler. This is not rare species in this locality.

70. *Dendroica cornata,* (Gray.)—Yellow-rumped Warbler. This is a common species in this locality.

71. *Dendroica Blackburniæ,* (Baird.)—Blackburnian Warbler. At some seasons this species is very abundant.

72. *Dendroica castanea,* (Baird.)—Bay breasted Warbler. Scarce, some seasons ; never rare.

73. *Dendroica pinus,* (Baird.)—Pine-creeping Warbler. This species is met with throughout the summer ; very probably breeds. Not a common species.

74. *Dendroica pennsylvanica,* (Baird.)—Chestnut sided Warbler. Usually scarce ; abundant, spring, 1867.

75. *Dendroica caerulea,* (Baird.)—Blue Warbler. None observed, as yet, in this locality.

76. *Dendroica striata,* (Baird.)—Black-poll Warbler. Met with every season ; remains only a short time.

77. *Dendroica aestiva,* (Baird.)—Yellow Warbler. Common summer residents.

78. *Dendroica maculosa,* (Baird)—Black and Yellow Warbler. Common during May.

79. *Dendroica tigrina,* (Baird.)—Cape May Warbler. I have not met with but one specimen of this species from this section.

80. *Dendroica palmarum,* (Baird)—Not met with as yet.

81. *Dendroica superciliosa,* (Baird.)—Yellow-throated Warbler. None observed.

82. *Myiodioctes mitraeus,* (Audubon.)—Hooded Warbler. None observed.

83. *Myiodioctes pusillus,* (Bonap.)—Green Black Cap Flycatcher. This is not an abundant species ; specimens are, however, to be met with every spring.

84. *Myiodioctes canadensis,* (Aud.)—Canada Flycatcher. A widely distributed and abundant species.

85. *Setophaga ruticilla,* (Swains.)—Red Start. This, the most beautiful of the Flycatchers, is a common and abundant species. It is to be met with throughout the summer, and probably breeds here.

86. *Tanagra rubra,* (Viell.)—Scarlet Tanager. One of the most beautiful summer visitors ; common.

87. *Hirunduo horreorum*, (Barton.)—Barn Swallow. This is one of the most abundant of the swallow tribe.

88. *Hirundo Lunifrous*, (Say.)—Cliff Swallow. A common summer visitant.

89. *Hirundo bicolor*, (Viell.)—White-bellied Swallow. Common summer visitant.

90. *Cotyle riparie*, (Boie·)—Bank Swallow, common summer resident.

91. *Cotyle serrepennis*, (Bonap.)—Rough-winged Swallow.

92. *Progne purpurea*, (Boie.)—Purple Martin. Resident throughout the summer.

93. *Amphelis Cedrorum*, (Baird.)—Cedar Bird. Wax-wing. Common resident all the year.

94. *Collyrio borealis*, (Baird.) - Great Northern Shrike. Butcher Bird. A northern species resident during winter.

95. *Collyrio chemungensis*.—Rufous Rumped Shrike. Like above, rump rufous. Variety.

96. *Viero olivaceus*, (Vieill.)—Red Eyed Viero. This is a common summer resident ; breeds with us.

97. *Viero gilvus*, (Bonap.)—Warbling Viero. One of the most abundant of the Flycatchers.

98. *Viero novaboracensis*, (Bonap.)—Not observed.

99. *Viero solitarius*, (Vieill.)—Blue Headed Flycatcher. This is an abundant species during summer.

100. *Viero flavifrons*, (Vieill.)—Yellow-throated Flycatcher. A common and abundant species during summer.

101. *Mimus polyglottus*, (Boie.)—Mocking Bird. I have never met with this species, but am creditably informed that a pair nested one season, a few miles south, in the State of Pennsylvania.

102. *Minus carolinensis*, (Gray.)—Cat Bird. This sweet songster is one of our most common summer residents.

103. *Harporhynchus rufus*, (Cab.)—Brown Thrasher. A common summer resident.

104. *Thriothomus bewickii*, (Bonap.)—Not observed.

105. *Cistothornus palustris*, (Cab.)—Long-billed Marsh Wren. A common species in marshy woods.

106. *Cistothornus stellaris*, (Cab.)—Short billed Marsh Wren.

107. *Troglodytes aedon*, (Vieill.)—House Wren. A common and abundant species.

108. *Troglodytes hymealis,* (Vieill.)—Winter Wren. Met with during winter in sheltered places, particularly among the drift wood along the river course.

108. *Certhia americana,* (Bonap.)—Brown Creeper. The nest of this species is built of dry twigs attached to the sides of some perpendicular object. I discovered one on the attic of a deserted log house ; the nest rested upon the inner projection of the gable clap-board, and was cemented together with a gummy or gelatinous substance. This is a common species remaining during the summer.

110. *Setta carolinonsis,* (Gmel.)—White bellied Nuthatch. Resident throughout the year ; common.

111. *Setta canadensis,* (Linn.)—Red-bellied Nuthatch. I have not met with this species in this section.

112. *Polioptila caerulea,* (Sclater.)—Blue Gray Clycatcher. Not yet observed.

113. *Lophophanes bicolor,* (Bonap.)—Crested Tit. Two specimens only of this species have come under my observation.

114. *Parus arlicapellus,* (Linn.)—Chick-a-dee. This well known species occurs throughout the state during the whole year. It is an active little bird, always busily engaged in seeking out its food, either upon the trunks of trees or snapping up winged insects, after the manner of the true Flycatchers It builds its nest in hollow trees, and lays from four to six eggs. Usually raises two broods in the season. P. carolinenis has not been detected.

115. *Eremophila cornuta,* (Boie.)—Shore Lark. This species has no been identified.

116. *Pinicola canadensis,* (Cab.)—Pine Grosbeak. This species has been met with in the state during severe winters. I have no knowledge, however, of its occurrence in this locality.

117. *Carpodacus purpureus,* (Gray.)—Purple Finch. This species is, I think, a constant resident. It has been met with at all seasons of the year.

118. *Chrysomitris tristis,* (Bonap.)—Gold Finch, Yellow Bird Resident throughout the year. Common.

119. *Chrysomitris pinus,* (Bonap.)—Pine Finch. This species has not been met with.

120. *Curverostra americana,* (Wils.)—Red Crossbill. This northern species visits us during winter. I have met with them as late as July, which leads me to believe they breed with us. They are not a common occurrence.

121. *Aegiothus linaria,* (Cab.)—Lesser, Red Poll. This gregarious

species is a northern visitor, and during some winters quite common. have a specimen which answers closely to the description of *A canescens*.

122 *Plectophanes nivalis*, (Meyer.)—White Snowbird. Snow Bunting. This northern species is to be met with every winter ; it is never abundant, however.

123. *Passercullus savanna*, (Bonap.)—Savanna Bunting. Common spring and fall on the low swampy grounds along the river.

124. *Poocaetes graminens*, (Baird.)—Bay-winged Bunting. Summer resident ; common.

125. *Coturniculus passerinus*, (Bonap.)—Yellow-winged Bunting. Resident nearly the whole year.

126. *Zonotrichia leuchophrys*, (Swains.)—White-crowned Sparrow. Resident during summer ; quite common.

127. *Zonotrichia albicollis*, (Bonap.)—White-throated Sparrow. Not so common as the above, but never rare.

128. *Junco hymeales*, (Sclater.)—Snow Bird. Resident throughout the year ; common.

129. *Spizella monticola*, (Baird.)—Tree Sparrow. Canada Bunting. Common winter resident. Prefers the open plains.

130. *Spizella pusilla*, (Bonap.)—Field Sparrow. Summer resident ; common.

131. *Spizella socialis*, (Bonap.)—Chipping Sparrow. Common.

132. *Melospiza melodia*, (Baird.)—Long-billed Sparrow. Summer resident ; abundant.

133. *Melospiza palustris*, (Baird.)—Swamp Sparrow. This shy retiring species is a common summer resident.

134. *Passerella ilica*, (Swains.)—Fox colored Sparrow. A gregarious species met with fall and spring. Breeds further north.

135. *Euspiza americana*, (Bonap.)—Black throated Bunting. Resident during summer.

136. *Guiraca ludoviceana*, (Swains.)—Rose-breasted Grosbeak. Resident during summer ; not common.

137. *Guiraca caerulae*, (Swains.)—Blue Grosbeck. Although this species has been met with in this state, and east as far as Maine, I have no knowledge of its having been obtained in this locality. Its migratory course seems to lay along the Atlantic coast.

138. *Cyanospiza cyanea.*—Indigo Bird. Resident during summer. Breeds. Young leave the nest as soon as July 4th.

139. *Pipilo erythropthalmus*, (Vieill.)—Chewink.   Towhe Bunting. Resident during summer.   Common.

140. *Dolichonyx oryzivora*, (Swains.)—Bobolink.   Arrives about the first of May ; abundant.

141. *Molothrus pecoris*, (Swains.)—Cow Black Bird.   This species remains with us nearly the whole year.   It builds no nest.   I found the young of this species in the nest of the Baltimore Oriole.   There were two young Orioles and one Cow Bunting in the nest.   The Cow Bunting was nearly large enough to fly, while the young Orioles were hardly fledged This was evidence that the intruder had received the lion's share of the daily food furnished by the mother.   There was one young Oriole dead under the tree.   I have also observed a pair of King Birds. (*Tyrannus carolinensis*,) feeding a young Cow Bird with their own brood ; and in this case it was remarkable to see the extra attention shown the supposititious bird, principally, I judged, on account of the clamor made on the arrival of a new supply of food.

142. *Sturnella magna*, (Swains.)—Meadow Lark.   Resident nearly the whole year.

143. *Icterus baltimore*, (Daub.)—Baltimore Oriole.   This beautiful bird is a constant resident during summer ; common.

144. *Icterus Spurius*, (Bonap.) Orchard Oriole.   Although occurring further east, I have no knowledge of its ever having been observed in this locality.

145. *Scolecophagus ferrugineus*, (Swains.) Rusty Blackbird.   Resident during summer ; common.

146. *Quiscalus Versicolor*, (Vieill.)—Crow Blackbird,   Resident during summer ; abundant.

147. *Corvus americanus*, (Aud.)—Common Crow.   Resident throughout the year.

148. *Zenaidura carolinensis*, (Bonap.) Carolina Turtle Dove.   Summer resident ; common.

149. *Bonesa umbellus*, (Stephens.)—Ruffled Grouse.   Pheasant constant resident ; common.

150. *Ortyx Virginianus*, (Bonap.)—Quail.   Resident throughout the year.

151. *Ardea herodias*, (Linn.)—Great Blue Heron.   Crane.   Common ; particularly during July and August.

152. *Ardetta exilis*, (Gray.)—Least Bittern.   Shy and solitary in its habits ; not rare.   Resident during summer.

153. *Botaurus Lentiginosus*, (Steph.)—Bittern.   Resident during summer.

154. *Butorides virescens*, (Bonap.)—Green Heron. Summer resident ; common.

155. *Charadrius virginicus*. (Borck.)—Golden Plover. Met with during spring and fall ; breeds further north.

156. *Aegialitis viciferrs*, (Linn.)—Killdeer. Resident during summer.

157. *Aegialitis semipalmatus*, (Bonap.)—King Plover. I have not been able to obtain a specimen of this species.

158. *Philohela minoro*, (Geml.) Woodcock ; resident nearly the whole year.

159. *Philohela minora*, (Gmel.) Woodcock. Common ; resident nearly the whole year.

160 *Gallinago wilsoni*, (Temm.)—English Snipe. Common. spring and fall.

161. *Macrorhamphus griseus*, (Leach.)—Red-breasted Sandpiper. This species is rare in this locality.

162. *Tringia alpina*, (Cassini.)—Red backed Sandpiper. This widely distributed species is of constant occurrence during spring.

163. *Tringia Maculata*, (Vieill.)—Jack-Snipe. I have not met with this speeies.

164. *Tringia wilsoni*, (Nuttall.)—Least Sandpiper. Ox-Eye. This species has not been detected in this locality ; but I have no doubt it will be met with.

165. *Gambetta melanoluca*, (Gm.) Bon. Tell Tale. This species is often met with during spring. I do not know that it breeds here.

166. *Gambetta flavipes*, (Bon.)—Yellow Legs. Common during spring and fall ; usually seen in small parties.

167. *Rhyacophilus solitarius*, (Wils.)—Bon.—Solitary Sandpiper. A common species during spring and fall.

168. *Tringoides macularius*, (Gray.) - Spotted Sandpiper. Tipup. The most abundant of the wading birds that visit us.

169. *Actiturus bartramius*, (Bonap.)—Field Plover. This is a common species with us.

170. *Rallus virginianus*, (Linn)—Virginia Rail. Not yet identified.

171. *Porzana carolina*, (Vielll.)—Common Rail. This is the only species of the Rail that I have met with in this locality.

172. *Fulica americana*, (Gmel.)—Coot Mud Hen. This species is of common occurrence during spring and fall.

173. *Gallinula galeata*, (Litch.)—Bon.—Florida Gallinule. I have a specimen of this species in my collection ; the only one I know from this locality.

174. *Barnicla canadensis*, (Boie.)—Canada Goose. This is a common species during spring and fall.

175. *Anas boschas*, (Linn.)—Mallard Duck. A common species, This is supposed to be the original of the domestic duck.

176. *Anas abscura*, (Gmel.)—Black Duck. This is a common species during spring and fall.

177. *Defila Acuta*, (Jaynes.)—Pintail Duck. I have not met with this species, but have no doubt it will be found here.

178. *Nettion Carolinensis*, (Baird.)—Green winged Teal. This species is of common occurrence, on our ponds and rivers.

179. *Querquedula discors*, (Steph.)—Blue winged Teal. This is a common occurring species.

180. *Spatula clypeata.* (Boie.)—Shoveller. This species is occasionally met with on our river, in the spring.

181. *Chaulelasmus streperus*, (Gray.)—Gadwell, Gray Duck. I have a specimen of this species, shot on New Town Creek.

182. *Marcca Americana*, (Steph.)—Baldpate ; American Widgeon. This species passes through our territory, on their annual migrations. Specimen obtained ; common.

183. *Aix sponsa*, (Boie.)—Wood Duck. This beautiful species is of common occurrence with us during summer.

184. *Fulica marila*, (Baird.)—Big Black ; Head Scaup Duck. This species has been met with.

185. *Fulix affinis*, (Baird.)—Blue Bill. This species is met with during spring and fall.

186. *Fulix collaris*, (Baird.)—Ring neck Duck. This species is to be met with at most all seasons of the year.

187. *Aythya americana*, (Bon.)—Red Head. This is one of the most common occurring species, in this locality.

188. *Bucephala Americana*, (Baird.)—Golden eye. Whistle wing. This is a common occurring species ; common to Europe and America.

189. *Bucephala albeola*, (Baird.)—Buffle Head. This is a common occurring species in this locality.

190. *Harelda glacialis*, (Leach.)—Old wife. Long tail. Not a common occurring species, I have the only specimen I have met in this locality.

191. *Erismatura rubida*, (Bonap.)—Ruddy Duck. I have not succeeded in obtaining a specimen of this species, although its occurrence here should not be rare.

192. *Aythya vallisnaria*,—Canvas Back Duck. Specimen in my possession.

193. *Mergus Americanus*, (Cassine.)—Goosander sheldrake. Fish Duck. This species is of common occurrence.

194. *Mergus sarrator*, (Linn.)—Redbreasted Merganser. This species is not often met with in the locolity.

195. *Lophodytes cucullatus*, (Rich.)—Hooded Merganser. This is not a common occurring species, yet specimens are often obtained.

196. *Larus Argentatus*, (Brun.)—Herring Gull, Winter Gull. This species is seen in our river every winter.

197. *Colymbus torquatus*, (Brun.)—Loon. Great Northern Diver. Common in our river in spring.

198. *Podiceps griseigena*, (Gray.)—Rednecked Grebe. This species is a winter visitor and often met with during fall and spring.

199. *Podiceps cristatus*, (Lath.)—Crested Grebe. Common during spring and fall.

200. *Podiceps cornuta*, (Lath.)—Horned Grebe. This species if of frequent occurrence during fall and winter.

201. *Podilymbus podiceps*, (Lawrence.)—Dipper Water Spirit. Common during fall and spring, on our rivers and large ponds.

DESCRIPTION of Capromys ingrahami.-- Discovered by D. P. Ingraham, a member of this Academy. By J. A. ALLEN.

Extracted by permission from Buel. Am. Mus. Nat. His., Vol. III, No. 2.

The Museum has recently received specimens of a small, short-tailed *Capromys*, collected by Mr. D. P. Ingraham on one of the Plana Keys, Bahama Islands, where he found the species abundant. Its nearest ally appears to be the *Capromys brachyurus* of Jamaica, a species (sometimes erroneously synonymized by authors* with the *Plagiodontia ædium* F. Cuvier of Hayti and San Domingo) which it resembles somewhat in external characters. It is also of interest to note in this connection that an allied form (*Capromys brachyurus thoracatus*) has been recently described by Mr. F. W. True from Little Swan Island, off the coast of Honduras.†

Capromys ingrahami, sp. nov.

Similar in size and proportions to *Capromys brachyurus* Hill from Jamaica.

Pelage coarse and harsh at the surface, softer beneath. General color above mixed reddish or yellowish brown, gray and black, giving the general effect of grayish brown. The hairs individually are plumbeous at base, passing into blackish, subapically ringed with pale yellowish gray varying to yellowish brown, the extreme tip blackish; with these however, are many hairs wholly blackish. Front and sides of head rather darker and with less brown; upper surface of both fore and hind feet brighter brown than the back; soles black, warty; basal half of tail well haired, pale rusty brown, as also the long hairs at its base, in contrast with the general coloration above; apical half blackish, the hairs short but concealing the annulations. Below, nearly uniform pale yellowish brown, barely a little more yellowish posteriorly. Ears of the usual form in *Capromys*, blackish, scantily haired on both surfaces, slightly fringed with

---

* *Cf.* Trouessart, Cat. Ram. viv. et foss. Fasc., III, p. 125.

† Proc. U. S. Nat. Mus., 1888, p. 469.

dusky hairs on the anterior border. Whiskers long, blackish. Thumb rudimentary, being little more than a tubercle armed with a very short nail.

Measurements.—Head and body (two specimens), 280 to 320 mm.; tail vertebræ (still *in situ* in one specimen). 55 ; hind foot, 53 to 55 ; fore foot 30 to 32 ; ear, height from crown, 16 ; greatest width, 17.

Skull.—Total length, 63 mm.; greatest width, 32 ; least width between orbits, 17·5 ; length of nasals, 21 ; length of frontals, 21 ; length of upper molar series, 15 ; distance between inner margins of the upper anterior molars, 2.5 ; distance between inner margins of posterior molars, 5.5 ; distance from anterior upper molars to incisors, 16 ; length of lower jaw (tip of incisors to posterior edge of condyle), 44 ; tip of incisors to tip of coronoid process, 48 ; height of condyle, 18.5 ; length of lower molar series, 14.5.

A second and somewhat older skull is slightly larger (total length, 66 mm.), but not otherwise different.

Type, No. ₁₀₄₄. Am. Mus. Hist., Plana Keys, Bahamas, Feb. 1891 ; D. P. Ingrahom.

Mr. Ingraham collected 20 specimens of this animal, 14 of which he has kindly transmitted to me for examination, together with his MS. notes on their habits. The skins are distorted, unfilled specimens, and hence unsatisfactory for study and measurement. The above description was based on two specimens kindly presented by Mr. Ingraham to the museum. The 12 additional specimens since received enable me to speak further of their external characters. The majority of the specimens present no very noteworthy differences from those above described. As nearly as can be judged they are nearly all of about the same size ; several are smaller and evidently (judging by the femora attached to the skins) younger than the others. Unfortunately the specimens are not labeled with the sex, and it is impossible to determine the sex from the skins. Mr. Ingraham assures me, however, that he noticed no sexual difference in either size or color.

As regards coloration, some are quite strongly suffused with yellowish and have less black ; others are faintly suffused, giving the general effect above of pale yellowish gray mixed with black. Below the color varies from soiled white faintly suffused with buff to rather strong buff. In most specimens the area surrounding the base of the tail, and also the basal third or half of the upper surface of the tail (sometimes the whole upper surface of the tail) is strongly rufescent brown ; in other specimens this region is nearly or quite concolor with the back. A single specimen is melanistic, being entirely brownish black, varying to nearly black over the medial line of the dorsal region, mixed with reddish brown, more especially over the lower back, sides of the body and feet.

Mr. Ingraham has kindly furnished the following notes respecting the discovery and habits of the animal it gives me pleasure to name in his honor, in recognition of his work as a collector during several winters spent by him in the Bahamas and in Yucatan. He says :

"On the morning of February 11, 1891, we anchored under the lee of the eastmost of the Plana Keys, in latitude about 22 deg. 33 min. north, longitude 72 deg. 30 min. west, and about half-way between the northeast point of Acklin Island and Mariguana of the Bahamas ; and on going on shore we saw unmistakable signs of the little rodent known among the natives as the 'Hootie' [=Hutia.]

"The key is a small rocky islet, the highest point of which is probably not more than fifty feet above the surrounding ocean, with crevices and caves worn in the rocks by the action of water, and in many places broken strata of rocks piled upon each other, leaving cracks and crevices between and beneath them. The islet may be slightly more than half a mile wide and four or five miles long, entirely without fresh water except in the rainy seaseon, when pools of fresh water may be found in the holes in the rocks. There is a small growth of shrubby bushes in the rocky crevices, and in some parts of the lower ground a growth of black button-wood, and on the western end of the islet a light growth of the silver-leaved palm, with here and there different forms of cacti scattered over the island. A few paw-paw trees were also found where the seeds had evidently been dropped. About a mile and a half west of the key is another small key, of about the same size and of the same geological formation, but separate from it by a deep passage. This is the only land within twenty miles or more, and my sailing charts indicate a depth of water of several thousand feet.

"Although these islands are only about a mile and a half

Skull of the Capromys ingrahami

As intimated at the beginning of this article, *Capromys ingra-hrmi* is most nearly allied to the Jamaican *C. brachyurus* Hill, so far as can be judged by descriptions, no specimens of it being known to be extant in this country. Through the kindness of Mr. F. W. True, Curator of Mammals at the U. S. National Museum, I am able to make a direct comparion of *U. ingrahamis* with his *Capromys brachyurus thoracatus*, from Little Swan Island, off the coast of Honduras, of which he has sent me the skull of the type and one or two skins on which the subspecies was based. In general effect the coloration of some of the Plana Keys specimens is indistinguishable from the skin of *C. b. thoracatus;* others, however, are more suffused with reddish brown. *C. ingrahami* is a much smaller animal, with a relatively longer tail. The skulls of the two forms show the same discrepancy in size, and besides differ much in various structural details. The anteobital portion of the skull in *C. ingrahami* is much smaller with the nasals much narrower and much less arched at the anterior border ; the anterior palatine foramen is one-half narrower and much shorter ; the maxillary portion of the zygomatic arch is much less expanded laterally, and the palatal surface is more deeply pitted in front of the molar series ; the palatine surface is less extended posteriorly, terminating on a line separating the third and fourth molariform teeth, instead of extending as far as the posterior border of the last molar. The most striking difference is seen in the form of the zygomatic arch, the malar being one-third less in vertical expansion, and lacking entirely the angular expansion at the lower posterior border, which forms so prominent a feature in *C. b. thoracatus* (see Figs. 9 and 10,) and also in *C. pilorides* and *C. prehensilis*. The two forms also differ slightly in details of dentition (see Figs. 1 to 4,) and in the form of the mandibular rami.

C. ingrahami appears to resemble *Plagiodontia œdium* of Hayti in size, coloration, and in its relatively short tail, but it being a true *Capromys* needs no further comparison with *Plagiodontia*.

The present is by no means the first record of *Capromys* from the Bahamas. Catesby's *Cuniculus bahamensis* is evidently one of the larger species of the genus, but which one, or whether really from the Bahamas, is at present beyond determination.

Columbus, however, on his first voyage to the West Indies evidently found some form of the genus abundant at nearly all of the several Bahama Islands he visited ; and Mr. C. B. Cory informs me that "a peculiar large rat, probably a *Capromys*," is said to occur abundantly on the island of Mariguana, a few miles to the eastward of Plana Keys. Mr. Ingraham, however, replying to my inquiries on this point, writes me that he spent from the 22nd of February to the 30th of March, 1891, at the island of Mariguana. He says : " The island has a coast line of about seventy-five or eighty miles, and I have walked nearly or quite half of this distance. I have been four or five miles into the interior, and indeed there is not a part fifteen miles in extent that I have not visited. I say no signs of *Capromys* anywhere on the island, nor did I hear of any such animal from the inhabitants, who, however, repeatedly told me of the ' Hootie' on the Plana Keys. Hence I' may say unhesitatingly that it is not found on the island of Mariguana."

Mr. Ingraham, who has visited a large number of the keys and islands of the Bahama group, further informs me that he has never heard of the existence of any similar animal elsewhere in the Bahamas. An animal so helpless and easily destroyed as the Hutia, may, however, have formerly existed at many points in the Bahamas and Antilles, where it is now extinct.

The first European explorers of the West Indies found these peculiar rat-like animals abundant in various parts of the Antilles, and vague descriptions of them were given under their various native names by the writers of the sixteenth and seventeenth centuries, notably by Oviedo in his " Historia general de las Indias," published in 1547, and later by Rochefort, Duturtre, and Browne. As these little beasts were in great quest as food, from the delicacy of their flesh, by both the natives and the Spanish colonists, they quickly began to become scarce, a fact noted even by Oviedo, who says they were hunted by dogs brought from Spain. They were so common in Jamaica at the time of Columbus' visit that he is said to have " victualled the famous canoe expedition of Diego Mendez with them."* The narrators of his voyage make frequent mention of their abund-

* Zool. Journ., Vol. IV, 1829, p. 277.

ance not only in the Bahamas and at Jamaica, but also in Cuba and Hispaniola. Oviedo speaks of three kinds, and later writers mentioned others, without, however, describing them so as to give a very clear conception of their characters. They have been referred to as occurring throughout the Greater Antilles, except in Porto Rico, and in the Bahamas. The earlier natural history compliers introduced them into their works, greatly to the distraction of later systematic writers.

Although these animals are apparently still not uncommon at certain localities on the larger islands, they have doubtless everywhere greatly decreased in number, and probably at many points have been wholly extirpated. Though said to be still common in Hayti and San Domingo, and in portions of Cuba, they have been practically exterminated in Jamaica. Specimens, however, are very rare in collections, and even at this late day our knowledge of the group is very inexact, while some of the forms have doubtless already become extinct.

Five or six species have of late years been commonly recognized, about as follows :

1. *Capromys pilorides* (Say.)   Hab. Cuba.

2. *Capromys prehensilis* Poeppig.   Hab. Cuba.

3. *Capromys brachyurus* Hill.   Hab. Jamaica.   Now nearly extinct.

4. *Capromys brachyurus thoracatus* True.   Hab. Little Swan Island, off coast of Honduras.

5. *Capromys ingrahami* (as above described.)   Hab. Easternmost of the Plana Keys, Bahamas.

6. *Plagiodontia œdium* F. Cuvier,   Hab. San Domingo.

Of doubtful status is *Capromys melanurus* Poey, from Cuba. Possibly some of the other supposed nominal species may have a valid basis, as I have before me three distinct species alleged to have come from Cuba. Besides, I am informed on credible authority that a long-tailed, as well as a short-tailed-Hutia is found in Hayti. Nothing, however, can be done in the way of a satisfactory revision of the group with the material at present accessible in American museums.

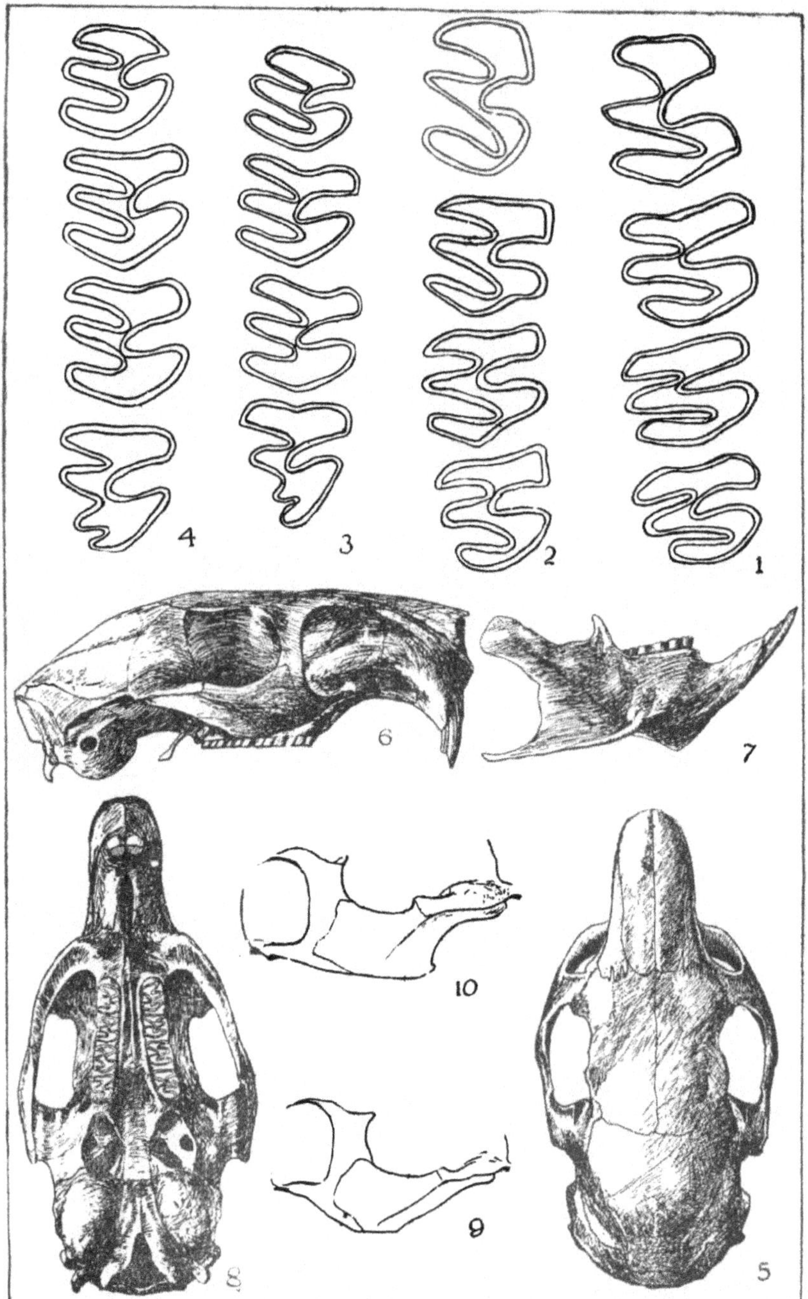
4
3
2
1
6
7
10
9
8
5

38

Explanation of the Figures.

Fig. 1. Pattern of enamel folds, left upper molar series, in *Capromys ingrahami* (No. ⁸⁰³⁸⁸, Am. Mus.)

" 2. Same, in *C. brachyurus thoracatus* True (No. ²¹⁸⁸¹, U. S. Nat. Mus.)

" 3. Same, left lower molar series, in *C. ingrahami*. From same specimen as Fig. 1.

" 4. Same, left lower molar series in *C. brachyurus thoracatus*. Same specimen as Fig. 3.

" 5-8. *C. ingrahami*. Figs. 5-7 from same specimens as Figs. 1 and 3 ; Fig. 8, from a somewhat older and larger specimen (No. ⁸⁰³⁸⁸, Am. Mus.,) No. ⁸⁰³⁸⁸ having the base of the skull imperfect.

" 9. Zygoma of *C. ingrahami* from same specimen as Figs. 1. 3, 5-7.

" 10. Zygoma of *C. brachyurus thoracatus* True, from same specimen as Figs. 2 and 4.

A NEW ELECTRIC Chronograph.---D. R. FORD, D. D.

Astronomy is a science of precision. It is founded upon exact measurements, both of time and space. It's apparatus must be of the most perfeôt mechanism, and is costly. The requirements of theoretical astronomy are far ahead of the artizans in brass and glass. Thus, the true determination of solar and planetary distances and magnitudes depends upon accurate circles, and longtitude depends upon exact measurement of time. The correct measurement of time into tenths or hundredths of a second is very difficult unless we make use of a Chronograph. By the old eye-and-ear method, an old and practiced astronomer might estimate the half or the fourth of a second approximately. But the observer was embarrassed by his method and his environment. To hold pencil and paper in the hands, to keep the eye upon a faint star, while counting and recording clock-beats, often in winter cold Transit Room, could not be very favorable to exactness.

These veterans of science had always known that there is in every observation an insidious source of error called the personal equation. But there existed for ages, no prime apparatus for eliminating this error.

The need of a Force and a Mechanism to meet these difficulties was great. Such a relief, they felt, would secure a much finer time record, would save much time and labor, would help eliminate personal equation, and would increase the number of good observations, while aiding to secure their accuracy.

The years 1848 and 1849 found several American astronomers and artizans engaged upon this problem. The old eye-and-ear methods were too slow and inaccurate. The circuit breaking clock was devised each in his own way by Bond, Saxton, Mitchell and Locke. Next came, during this period, the electric recording apparatus, simply a Morse telegraph fillet at first, then afterwards the rotating disk, and the revolving cylinder, equipped with either the pen, or stylus, actuated by the clock magnet. Of all these forms, our subsequent experience shows great favor to that of Bond.

M'CONNELL'S ELECTRIC CHRONOGRAPH.

But probably one of the earliest successful attempts, in this country, to construct and exhibit a Chronograph for astronomical purposes was that of Professor O. M. Mitchell, astronomer, and afterward a General of good record in the Union armies. Those members who were present at the Meeting of the American Association for the advancement of Science, at New Haven, Ct., in 1849, will probably remember the lively discussion of Prof. Mitchell's paper announcing his invention of such a mechanism. It was, however, a success. The first form used by him, was that of a circular, rotating, horizontal disc, over which stood two styluses, one of which was actuated by an electro-magnet, and an astronomical clock, while the other was made to record the instant of a star transit, by means of an electric key in the hand of the observer at the telescope. Each stylus made up the record by imprinting dots along a spiral line always approaching the center of the paper-covered, rotating disc.

The force which moved the revolving disc was a heavy weight, made to actuate a train of wheels. The greatest trouble in this early form of chronograph was to secure an absolutely uniform motion. In truth, every form of automatic recording instrument is exposed to this minute source of error. Those of Bond, Airey, and others, have been eminently successful in eliminating it, either by a spring-governor escapement like Bond's, a peculiar pendulum like Airey's or some form of fan, or ball governor more or less effective. All of these devices are somewhat expensive. The motive power has generally been that of gravity.

Whether the recording paper has been attached to a revolving disc or to a cylinder, whether the record signals have been traced by a metal stylus, or by a hollow glass pen holding some free flowing ink, they are necessarily costly, by reason of their complicated structure. This has restricted somewhat their use, except in well endowed observatories.

Some little time ago it occurred to Augustus MacConnell a member of the Elmira Academy of Sciences to construct a more simple chronograph of accurate workmanship, and at a cheaper rate of cost, if possible. The recording cylinder and hollow glass pen are the same form as in many others. But the motive and regulating power is an electric current from a battery of a half dozen Lelance cells of extreme cheapness, acting on open circuit. This battery in our Observatory has been cleaned but once in three years and it never freezes. The recording pen is actuated by the same circuit. This self-regulating motor may also be used upon any electro-magnetic machine where accuracy of movement is required. Its merit of uniformity when used in a chronograph is very marked. By its use the chronograph really requires but one toothed wheel cutting into an endless screw upon the motor shaft. Two or more wheels may be used if occasion requires. "The nature of this invention consists in the combination, with a rotary electro-magnetic motor, of a conical pendulum revolving with the shaft of the motor and a suitable circuit-closer so arranged and constructed, that on any increase of the amplitude of the rotation of the conical pendulum,

(we use a seconds pendulum) beyond a certain limit, owing to an increase of the rotation of the motor, an interruption in the succession of currents through magnet of the motor will be caused, and its rate of rotation be consequently reduced until the desired limit is reached."

Of course a small rheostat must be included in the circuit to furnish a suitable resistance in keeping the battery power constant. If from imperfect mechanism, a slight variation from uniformity should occur, then a chronometer suitably joined in the circuit will secure a uniform rate with great certainty. The cost of constructing such a chronograph, whose cylinder will contain the records of three hours work among the stars, need not be more than one-third to one-half of the price heretofore paid. The record sheet is very uniform, neat and accurate, and being made with free-flowing carmine ink, it does not clog or freeze. The accompanying diagram may render aid in understanding the construction.

WHERE TO FIND Living Objects for Examination by Microscope.

It is of great advantage to the student to know the exact locality in which the different species in quest may be found, because the plant or animal which he seeks is often, nearly, if not quite invisible to the naked eye, and he must go home to his microscope, carrying simply a pail of muddy water, without knowing whether it contains anything of interest or not.

Perhaps the first object to attract the attention of the young microscopist in this region is the beautiful volvox globator. A live-box filled with these green globes, rolling and tumbling in every direction, never fails to delight the most indifferent observer. Elmira is particularly favored in its possession of the volvox globator, and even here I know but one place in which it is found, and that is a pond on the east side of Sullivan street, north of the woolen mill. It is called the 'old brick pond', not Mr. Weyer's brick pond in the brick yard, but an old deserted pond in a meadow, given over in the summer to blue flag and rushes, white arrow-head, and a tangle of marsh forget-me-nots. There are several ponds in the meadow, but one the largest, and of that only the south-east corner contains the volvox, or much else of interest. In the spring they are very rare, and it is the task of a morning to collect enough for an exhibit. I use in collecting a small, wide-mouthed bottle, wired to the end of a walking stick. The bottle is lowered gently, mouth downward, and brought cautiously to the right spot, it is then inverted, the air rushes out and all light objects are carried into the bottle by the force of the inrushing water, I then hold the bottle against the light, and if even one or two green flecks can be seen, empty its contents into a pail and dip again for more. Late in August or early in September, if the season has been propitious, and the pond has been well warmed, and also evaporated to small dimensions, hundreds of volvox may be taken at a single dip. Cold and rainy seasons are unfavorable to their development, and for several years these dainty globes have been very rare. Their entire disappearance I fear is not far off, as the pond I learn is to be soon filled up.

This same pond contains many crusstacea, fine large specimens of cyclops quadricornus among them, and different species of daphnia. It is also rich in infusoria, including paramacium, vorticella, stentor and others. The rotifer which the late Dr. UpDeGraff discovered, and named after Dr. Gleason, Anuroea Gleasonii lives here in great abundance. I believe however that Dr. UpDeGraff obtained his specimens from Sly's pond, in the Fifth ward. Leeches of all sizes, tadpoles and the young larvae of the dragon-fly (Libellula) also abounds in the Brick pond.

If one is in need of rhizopods, he will expedite matters by going directly to the Water Cure glen. There he must tear from the rocks the long silky masses of green algae and wring them in his pail. When he goes home he will find by dipping lightly with a pipette into the sediment at the bottom of his pail nearly all the beautiful shelled rhizopods pictured by Prof. Leidy in his great work on the "Rhizopods of North America."

Speaking of rhizopods I am uncertain whether or not it is generally known that many varieties of foraminifera shells may be obtained in perfect preservation from a new sponge. The sponge needs only be torn open and its contents dusted out upon a slide.

Sometimes late in the fall the algae in the Water Cure glen has contained quantities of amoeba proteus, a dozen times larger and more active than any, one can cultivate in an infusion. Under the microscope they look as large and luscious as oysters, and they certainly change their shape as rapidly as the old men of the sea. Diatoms and desmids are also numerous in the water wrung from this algae. The first diatoms which were found in this valley in sufficient numbers to warrant cleaning, I discovered by accident in wringing out the algae from the rocks in Rorick's glen, up the river. It was from this exhaustless supply that Dr. UpDeGraff prepared his beautiful slides of Chemung diatoms.

In the cold streams of the two glens above mentioned, there is a small larval salamander. Those about an inch long have large bushy gills at the sides of the head and are the most convenient creatures I know of, for showing the circulation of the blood. A frog is tedious to tie in position, and is always twisting out of focus. The necturous is more obstreperous still, and moreover is very rare in this valley. I believe there are a few in a spring two miles over east hill. But the young salamanders are easily caught and will live for months in an aquarium. At a moment's notice one can be transferred to a live-box, filled with water, the cover of which is pressed just tight enough to restrain all rebellious wrigglings, and the circulation in his gills and toes, and even the beating of his heart, examined at one's leisure. One may then have the satisfaction of returning the little fellow to the aquarium unharmed.

Fresh water polyps or hydra are not numerous in our waters. If one is desirous of studying their habits I know of but one way of obtaining a sufficient supply. I carry home a small pail full of brick pond water, containing floating duck-week. The water is empted into any glass aquarium and placed in a sunny window. If on the duck-week there is even one hydra, so rapidly will it multiply by budding, that in less than a week the side of the aquarium next the window will be covered with single hydras and small colonies. The entomostraca contained in the pond water will furnish them with food for a long time. I have obtained in this way both hydra viridis and hydra fusca.

There is considerable fresh-water sponge to be found in the Chemung river. In the fall a dip among the weeds by the bank often contains the staloblasts of a common species, pectinatella magnifica. The slow stream which flows through Gen. Diven's grounds, north of the city, is a rich field for the collection of diatoms, desmids and rhiropods.

Nearly all the places mentioned are inaccessible in winter, but across the road from the Brick pond is a clear spring which never freezes. Under the leaves of the water-cress which chokes its throat, live millions of gammarus polex. I have found no difficulty in obtaining all I wished at any time of the year.

There are as many methods of collecting as there are collectors. Dr. Gleason uses a small tin dipper attached to the end of a fishing rod, or else he puts on high rubber boots and wading into the stream wings out the weeds into his bottle. He prefers wide-mouthed bottles to a tin pail.

A gentleman who made a business of collecting living objects for the microscope, and mailing them to all parts of the country attended the meeting of the American Society, held here a few years since. He naturally wanted to find something to exhibit at the "soiree" and was directed to the Brick pond, but it seems never reached it. In the evening I was astonished at the large active colony of vorticella which he had under his microscope. He said that in a small pool left in the corner of a field, near Sullivan street, a little twig floating on the surface attracted his attention. He fished it out with his cane, and examining it with a powerful pocket lens, recognized the vorticello, put it into a bottle in his pocket and returned. It was a beautiful sight. As many as fifty of these living bell-flowers must have been in the field at once, all attached by their stems to the little twig. Some were expanded with their cillia thrown out, and in rapid vibration, others were snapping back on their spinal stems. while still others floated slowly out again into position.

There are always ways and means. One need never leave a valuable find because unprepared to carry it. Large leaves are very convenient, and a pocket handkerchief is always ready. A pocket handkerchief will carry anything from algae to salamanders, a frog or even a crayfish. To be sure one must have a good memory. It is sometimes embarrassing to discover a salamander in one's handkerchief. but what is one to do when he must have a salamander.

ANNA M STUART.

THE TIOGA RIVER and its Tributaries in connection with the great flood of June 1st, 1889.

At a meeting of the Academy of Sciences, held in June, a communication was read from Mr. Francis Collingwood calling attention to the recent extraordinary flood in the Chemung river on June 1st, and that he deemed the subject one which this society should take up, and secure while fresh in remembrance, such interesting data as it could, relating to the immense flow of water.

The Academy adopted the suggestion, and issued a circular with a number of questions which it was desirous to have answered. This circular was put in the hands of Mr. Jabin Secor, the Chairman of *The Meteorological Section,* and myself as Secretary, to have it distributed into the adjoining country lying along the course of the Chemung river and its tributaries.

The replies to the questions are in the possession of Mr. Secor, and he will make due report upon them. For my own part I have sought out—though with very imperfect results, some facts with reference to the Chemung tributaries which lie within the state of Pennsylvania. It is a very popular idea that the Chemung river has its rise in the counties of Steuben and Allegany, in this state. Such is not a fact. The really head waters of the Chemung are the Tioga and Cowanesque rivers, rising in the hills of Pennsylvania, in the two counties of Bradford and Tioga of that state.

First I will speak of the peculiarities of the country in Tioga county. Its area is about 1,125 square miles. Three mountain ranges penetrate into and two of them pass through the county in a direction about N 60° E. The southwest corner of the county is part of the general Potter-Lycoming-Alleghany mountain plateau cut through to its base by the canons of Pine Creek, and is drained south-westerly along its center line by the second fork (Babb's creek) of Pine creek, and the extraordinary spectacle is here exhibited of several large streams from the Wellsboro valley, flowing *towards* the north face of the mountain, *entering* in and uniting with the main stream along its middle line.

This topographical phenomenon is repeated in the next mountain range to the north and is an example on a small scale of a law illustrated more grandly by the rivers of the state of Ohio. The eastern end of this first mountain range of the Tioga county is a deep coal basin drained by the Tioga river, rising in the northern slope near Blossburg, and flowing north into New York state. This is the highland of the county, a thousand feet or more above the broad valley of Mansfield and Wellsboro to the northwest and the open rolling country of Bradford county to the north and east.

The second mountain range of Tioga county is a projection from Potter county through Shippen, Middleburgh and Tioga townships, in the latter of which it ends, as the afore mentioned Blossburg range. The northern slope drains its water into Crooked river, first easterly and then north-easterly into the Tioga river at Tioga village. Were the county divided centrally each way, into four equal squares, we should find the south-western square discharging all its waters into Pine creek flowing to the southwest and the other three squares discharging into Tioga river which empties into the Chemung river.

The third mountain range passes through the north-west corner of the county and from it descend the branches of Cowanesque river, which flows in a pretty straight line about N. 75 E., for 15 (?) miles at its foot. Having thus spoken of the mountain ranges of Tioga county, we may pass on to speak of its rivers, notably first of which the Tioga leaves the county and enters New York state at an elevation of nearly 1,000 feet tide, for the railroad grade at Lawrenceville on the state line is 1006 feet; at Mitchell's creek, 1,022 feet; at Tioga village, 1,042 feet; at Mill creek mouth, 1,077 feet; at Lambs' creek, 1,111 feet; at Mansfield, 1,163 feet; at Covington, 1,208 feet; at Blossburg, 1,348 feet. The Tioga descends threfore about 350 feet from Blossburg to Lawrenceville in a distance of 22 miles, in a nearly straight line (25 miles by its bends) at a rate of about 22 feet per mile for the first 9 miles and 11 feet for the last 14 miles.

From Blossburg 6 miles to the Fall Brook Coal Company's mines the fall is 500 feet (1,842 feet above tide) and from the mountain summit back of the mines several hundred feet higher.

Crooked creek rises at Little Marsh, the Antrim railroad terminus at 1,852 feet above tide. Antrim mines, 1,672 feet ; Morris Run terminus, 1,678 ; at Wellsboro on the divide between the waters which flow four ways 1,317 feet ; at Niles Valley, 1,179 feet ; at Hollidaystown, 1,151 feet. The mountains of Tioga county rise to a pretty general level of 2,000 feet above tide, and the broad valleys between their surfaces about 1,200 or 1,300 feet., The main water channels are cut down sharply to a depth of 1,000 feet or less. Inside the oblong oval basin of the Blossburg coal fields the Tioga river is joined by South creek, Fall Brook, Carpenter's Run, Taylor's Run, Morris Run, Coal Run, Johnson's creek and East creek ; all rapid streams; descending with the dip from the oval rim of the mountain and cutting deep furrow-like vales.

Cowanesque river—The valley of this river lies between the Mill creek basin in the south and the Cowanesque mountain basin on the north, and extends for 25 miles from the Potter county line to the Tioga river. Its breadth varies from 6 miles at its western to 10 miles at its eastern end. The south branches of the river, namely, Mill creek, Potters' brook, the Jameison, &c., drain its west end northward ; in the middle region small streams flow north. The eastern end is drained by the Elkhorn east, south-eastward into the Tioga at Tioga village. The northern tributaries are Potter brook, North Fork, Troups' creek. Holden brook and Camp creek.

Tioga county lies in the rain belt of 40 in, and it is indeed here fully that amount. Dry north-west winds favor radiation and evaporation. The rocks alternately expanded and contracted, are prepared for absorbing moisture from rains, so that every surface, hillslope and mountain steep are depositories of water till they overflow in excessive storms to devastate the valleys below. Thus I have sketched rapidly the Pennsylvania tributaries of the Chemung. The particularly local characteristic of our memorable flood has been briefly prefaced, preparatory to what additional work may be considered proper by this Academy. The amount of rainfall at this particular date, May 30th and 31st, and the rapidity with which it fell must be obtained from the papers to follow. One peculiar incident is this, that no rain of any consequence fell to the north-east and east of Elmira. At Ithaca I am told that the streams were hardly roiled.

Should we hear of severe and lasting rains in Tioga county, Pennsylvania, we may well feel more alarmed at the prospects of a flood in our Chemung than when we are told of the like rains in Steuben and Alleghany counties. Their streams are slower of descent and the valleys broader with more vegetation to hold back the rainfall moisture upon them.

ROBERT A. HALL.

The following "circular" letter was sent out into Chemung and Steuben counties in New York and Tioga county in Pennsylvania, for the purpose of gathering all the statistics possible relating to the June flood of 1889.

SIR—The Elmira Academy of Sciences is desirous of investigating the great flood of June 1st, in the Chemung river, and think you can perhaps help it. Will you be kind enough to answer the following questions, to the best of your knowledge, and return.

1. What day and hour did the rain begin ?
2. When was it heaviest ?
3. How long did the heavy rain last, and when did it stop raining ?
4. Have you any measure of the amount of water that fell, if so how much ?
5. In which direction was the wind ?
6. Was the ground well soaked with rain before this, or was it dry ?
7. When did the river or creek begin to rise at your place ?
8. How rapidly did it rise ?
9. At what hour was it the highest ?
10. Do you know the velocity of the water at the highest flood ?
11. Was there anywhere near you any reservoir or dam destroyed that would increase the height of the flood ?
12. Was the water ever as high or higher before, if so when ?
13. Can you draw a tolerably correct map of the river or creek (on a scale of 1-2 inch to the mile) that passes your place, from its source to

its mouth, giving the descent in feet and showing the gullies and smaller creeks that empty into it ?

Please give any other facts bearing on the subject of which you may have knowledge.

I enclose a letter, that was read before the Academy, written by Mr. Francis Collingwood, a civil engineer, an old-time member who has since become renowned in the engineering world. Its importance was recognized and an appropriation was made for the necessary expenses incident to carrying out a system of questions to be embodied in a circular to be sent out by the Academy to as many persons as could furnish replies to any or all of the questions asked.

Parties residing in the section of country drained by the Chemung will confer a great favor by making note of whatever data they may have, or can obtain, and forward the same to the Academy of Sciences. In due time this will be compared and the results will be furnished for the information of the public.

An early answer will be most thankfully received.

Yours most respectfully,
JABIN A. SECOR,
Chairman Section on Meterology.

Letter of Mr. Francis E. Collingwood, C. E., to the Academy.

ELMIRA, June 12, 1889.

Prof. D. R. Ford:

DEAR FRIEND—Learning to-day from you that the Academy of Sciences is to hold a session to-morrow evening, I feel impelled to write you a note for presentation to the meeting, claiming the right to do so as an old member.

I have been greatly impressed in the short time of my present visit, by the urgent need there is of a scientific examination into all the circumstances attending the recent flood ; and the Academy can render an eminent service to the community by taking the matter up and giving it a thorough consideration.

It is now becoming evident to all sanitarians that averages, while valuable, may be very deceptive, and for many reasons engineers and others find it desirable to know not only how much rain fell in a day, or throughout a storm, but how much fell in the period of greatest precipitation in ten minutes, in an hour, etc. Also where was the greatest precipitation, and what area did it extend over. For example, in an extraordinary storm in Connecticut (I think it was, as I write only from memory) there was found to be a very limited area, where from seven to nine inches of rain fell, as I now remember. Around this were zones successively of less and less amount down to that of a very moderate rainfall.

To come to the present case, the flood of 1865, I believe it was, which was the greatest known previous to this one, was undoubtedly magnified by the breaking of the state dam at Corning. This gave rise to an extraordinary wave, which passed in a few hours, but, with such a steep slope on its point as to give a velocity when going by the city of about ten miles an hour, as I roughly measured it.

As nearly as I can learn, the recent flood, while over-topping that by some eighteen inches, showed if anything a more uniform rise at the last. In both cases the breaking of embankments modify the results.

Now the questions arising are many, but we may put them down somewhat as follows :

First—Was there anywhere any dam or reservoir destroyed to cause the great increase in height reached by the water ? (I have heard of none.)

Second—The rain was not sufficient to explain the phenomena. (a). Where was the center of rainfall ? (b). What amount fell then, both total and in detail ? (c). The relation of surrounding areas contributing to the result ? (d). The extent and declivity of water-shed contributing ?

Third—The question of date and time at each point of observation recorded.

Fourth—Accompanying phenomena of direction of wind and storm, and other things required to give a ciear conception of the whole.

Fifth—If such were taken at any point, the velocity of the water at the highest flood is a very important item.

If we had all these facts it would be possible to answer, within

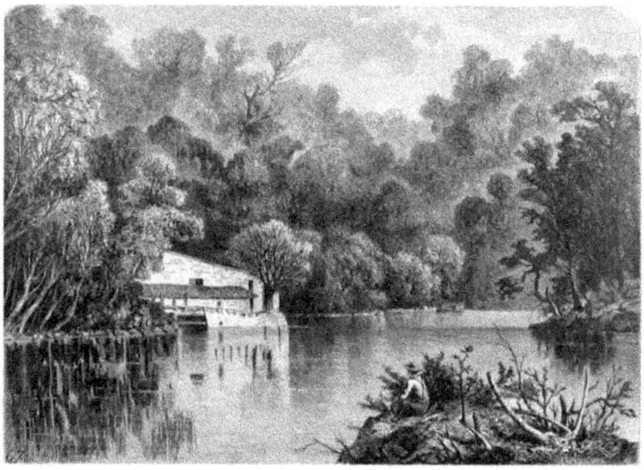

REPORT of the Special Committee of the Meteorological Section.

SUBJECT—CHEMUNG RIVER FLOOD OF JUNE 1ST, 1889.

A memorable storm was that which swept over our country, beginning at about 4 p. m., of Thursday, May 30, 1889. As a consequence we had what is known as the great flood of June 1st, 1889. In order to learn all that was possible concerning the storm, a resolution was passed by the Academy on June 13, directing a set of questions to be prepared and sent out asking for information. See page 50.

One hundred and fifty circular letters were mailed to persons living in the territory drained by the Chemung river system.

This system includes nearly the whole of Chemung, Steuben and smaller parts of Alleghany and Tioga counties in New York state, together with parts of Bradford, Tioga and Potter counties in Pennsylvania.

The Chemung river is formed by the junction of the Tioga and Conhocton rivers near Painted Post, in New York state, and flows for forty miles in a south-east direction, emptying into the Susquehanna at old historical Tioga Point, near the present village of Athens, in Pennsylvania. The Tioga river rises in Pennsylvania, in the Alleghany range, and is about sixty miles in length, its general direction is north-east, and its largest acquisition is the Canisteo river, which has its source in western Steuben and eastern Alleghany counties, and is fifty miles long ; its next largest branch is the Cowanesque river, which rises in the Alleghanies, and is fed from the highest watershed east of the Rocky mountains. The Conhocton river is fifty miles long and drains the rocky hills of northern Steuben.

From the reports sent in answer to our questions, we learned that the storm began during the afternoon of Thursday, May 30, and continued through Friday night, it was the heaviest between 10 p. m., and 3 or 4 a. m., of Friday night, and stopped about 6 a. m., of Saturday. In this city it did not cease until a little after 7 a. m.

Only a few correspondents had any record of the amounts of

water which fell. During the thirty-nine hours from 4 p. m., of Thursday, May 30th, to 7 p. m., of Saturday, June 1st, the rainfall in this city amounted to 2.92 inches, which is quite remarkable ; the rainfall for the entire month of May, with the exception of the last two days being 2.09 inches, it is evident that we did not get the greatest precipition, for the signal service observer at South Canisteo reports 6.25 inches for the same thirty-nine hours.

As to the course and general severity of the storm, Lieutenant Finley, of the signal service at Washington, give the best account, he said, " On May 27th, a storm started in Utah and " seemed to be quite developed in South California and the " whole Pacific coast. Thence it moved in a course made up of " ups and downs, which we call a sinous curve, on toward the " east, but making very slow progress on account of the high " pressure that prevailed on the Atlantic coast. When it struck " Ohio a trough-like vortex was formed, with its centre first in " Kentucky then in Ohio, and finally in Pennsylvania, where " the precipitation was the greatest. It was not a tornado, how- " ever, although the currents of air came rushing in toward the " centre from the south with a very high, and from the north " with a low temperature. This has ample demonstration in the " fact that the mercury in New York state and in Canada just " previous to the storm fell below the freezing point. In a " narrow trough the cold air could get much nearer the warm " than under any other condition. Therefore the gradient or " contrast of temperature was the greater, and the downpour of " the rain it caused was almost unprecedented.

" These were thus two main reasons for this great storm. " First, the very slow movement of the storm eastward on ac- " count of the high pressure in New England and Canada, for " had it come rapidly the rain-clouds would soon have been dis- " sipated ; Secondly, because the trough-like depression allowed " a more thorough mixing of the cold air that rushed in. As is " general the precipitation of vapor from moisture-laden air, as " dew, rain or snow, occurs where the atmosphere has been cool- " ed below the limit of vapor saturation. The trough then " brought about the conditions for a more than ordinary cooling " off of the rain-clouds, whose progress was abnormally slow.

" Hence the great precipitation. It was no cloud burst. There
" was no fall of a sheet of water from the heavens, but merely a
" very heavy rain storm in a hilly country.

" The water gathered on the sides of high ridges and concen-
" trated into creeks or dry runs in such quantities that the ordi-
" nary explanations failed to satisfy some minds. These had
" then to call out of their imaginations either the supernatural
" or a provision of normal action of the forces of nature.

" Now as to what became of the storm, the center of which
" was near Pittsburg, on May 30th. It divided, a part almost
" stationary over Lake Erie and finally moving slowly eastward.
" The other division developed as a secondary depression over
" North Carolina after which it took a northeasterly course over
" Maryland, New Jersey and New England, meeting the other
" portion of the storm Saturday, in Maine, where they had a
" heavy pricipitation."

Scientific investigations have explained the causes, the length,
breadth and force as a whole in miles, but the location as stated
by Lieutenant Finley is not to be construed as definite, as for
example, where he says, " The center of which was near Pittsburg," he does not mean within a mile or two of that city, for
Johnstown is nearly eighty miles distant, and the Cannemaugh
lake some fifteen miles further away.

The last great flood was the one of March 17, 1865 ; at that
time the river here was obstructed by an island nearly three-
fourths of a mile in length, now virtually gone. All reports
agree that there was more water on June first than then. At
the time of the June flood, the river began to rise about 5:30 a.
m., and continued for about fourteen hours, until about 7:30 p.
m. Coming up in that time seventeen feet. The rise up till
noon came from middle Steuben, the northern limit of the storm
area ; by the way of the Conhocton and Canisteo rivers. The
waters from among the Alleghanies by the way of the Tioga and
Cowanesque rivers were longer in reaching us and occassioned
the rapid rise from 3 to 6 p. m.

Mr. Robert A. Hall has prepared and presented this same subject in a paper showing the origin of this flood in the watershed
of the Pennsylvania district, in which lie the Tioga river and its
tributaries.

JABIN A. SECOR.

THE Chemung County Flora---Its relation to that of the Southern Tier Counties, and a brief comparison with other portions of New York State.---By T. F. LUCY, M. D.

On looking at a map of New York we shall find little Chemung almost the smallest in size of all the counties of the state, lying just north of the Pennsylvania boundary line, among what are generally known as the Southern Tier Counties. Thus it occupies the southernmost position of the western central counties of the state, almost square in outline, except a little jog in its northern boundary belonging to Schuyler County.

At a rough estimate from the maps at my disposal, I find its western boundary line on the seventy-seventh degree of longitude, west from Greenwich, being about twenty-eight miles from north to south, and its average width about twenty-two and a half miles, its area thus comprising something like five hundred and ninety-four square miles, the eastern boundary line following mainly the contour of Cayuta creek. Without going into any details as to its geological condition, I shall only mention its general topography. Also being at present unable to give any strictly accurate figures as to its highest elevation, Wellsburg being estimated at 850 feet above tide water, the hills of the county will probably reach an elevation of 1500 feet and perhaps higher, the general trend of the surface is from north to south, with an eastward dip as shown by the course of the Chemung river which enters the county at the southwest corner of the town of Big Flats, and flowing southeasterly across the county leaves it between Chemung and Waverly at about the Pennsylvania boundary line, and this sweep of the Chemung valley gives to our county its additional floral interest, for to this region of the county the young botanist may look for species not very likely to be found elsewhere in our limits, except it may be the more common species to be found in the northward valley of the town of Horseheads, and some along the main creek bottoms.

Our Chemung flora is essentially an inland one, and thus more uniform, and curtailed in the number of its species, being mainly that of an upland and alluvial flora, there being few ponds in the county. Evidently the primitive flora of Eldridge lake has been almost completely destroyed by the park there made, and it probably was at one time a station for species not found there now. Miller's Pond in the town of Southport, and Lowe's Pond in the town of Big Flats are the other two which have been visited botanically, the former well known for its white pond lily, *Castalia odorata*, and the latter has yet been only slightly explored. Judging from the numerous swampy spots now being drained and cultivated, there was at one time in the past many more than there are now. As far as I can estimate for the present, being without any data to govern my calculation, the highest part of the county seems to be in the town of Erin. There is also a good elevation in the southwestern corner of the town of Southport, where you leave Seeley creek and take the road over the hill, where near the Caton line is to be found another swamp whose centre is inaccessible. Lowman's swamp in the town of Chemung is one of the most interesting stations for some species that I have yet found in the county, and the one that has been most thoroughly examined, its flora is rich and varied, being cool and well shaded. The drainage of the county is as follows: first in the northwest corner of Catlin township the head waters of Post creek partially follow its western boundary line until it enters Steuben county, then Sing Sing creek flowing south empties into the Chemung river draining the township of Big Flats, next easterly comes Newtown creek, the largest in the county, through the northward trend of the valley, draining the townships partly of Horseheads and Elmira. Between these two in the southeasterly course of the Chemung, above its north banks rise the rocky slopes near the old mountain house, the home of a few rare species. Next comes Baldwin creek, enclosing a range of hills between itself and the Chemung in its lower trail, where the Sullivan monument now is, this hill having a rich upland flora. Eastward yet comes Wynkoop creek, flowing through Chemung township, the town of Van Etten in the northeast corner being drained by Cayuta creek which runs through nearly the southern centre of the township. Catharine creek flowing southward through the centre of Veteran turns sharp to the west then sends its waters northward into Seneca lake, the two streams only a mile or so apart in this portion of their respective courses. Newtown creek seeks the Atlantic ocean through the Chemung. Lastly Seeley creek running east and northeast drains the townships of Southport and Ashland. Thus we see the lowest part of the county is the valley of the Chemung, where we shall find a flora varying in several species from the other portions of the county, this flora being continued through Tioga and Broome counties with some additions, especially in Broome county, as a following comparison will show, for the conditions governing a flora locally seem to be rather the nature of the soil and its elevation rather than a few degrees of latitude and longitude, in a state flora some species seeking the highlands and others only following the alluvial bed of a large stream, or in other words, each species seeks and flourishes best in its most suitable environment, in conformaty to the law that in a previous paper I have formulated *That anatomy or structure is subservient to function and modified by the environment of the type.* Thus we find gaps of country between rare species over which the environment is unsuitable to the requirements of the plant or animal, and it will be ever thus with some species of either a flora or a fauna, some plants like some animals clinging only to the most suitable localities for their growth and the perpetuation of their kind. The flora of Chemung county may therefore for our present purpose be divided into five sub-floras.

The Chemung river with its alluvial flats. This comprises the whole length of the river in its course through the county, including the species occupying the narrows of Wellsburg, Chemung and along the north bank of the river for one or two miles near the Mountain House, including all the alluvial flats of Newtown Creek extending about north and south through the eastern central portion of the county to its northern limit, containing essentially an alluvial flora not found on the hill ranges. Here can be found some of the rarer species, such as *Ranunculus ambigens*, at the creek boundary line just south of the Village of Wellsburg. *Arisaema Dracontium*, once found in the town of Ashland on the river bank, and curiously enough this rare species in the county, after the great flood of 1889 is now growing in a flower bed in front of my present residence, having come down there from some station above the city where it has been vainly searched for many times ; *Polymnia Canadensis*, at the the Chemung narrows, above the Village of Chemung. *Pycnanthemum incanum*, above the Mountain House with that rarer plant yet *Potentilla arguta*. Also *Pycnanthemum lanceolatum* in Ashland and on Newtown Creek at Horseheads, where was found also, *Poterium Canadense*, *Lophanthus scrophulariaefolius* near Rorick's Glen, *Hydrangea arborescens*, found at the Wellsburg narrows in 1879. This interesting species was first seen by myself in 1864 in a cool shaded ravine south of Corning, Steuben co. Have since found it north of the Chemung, between Wellsburg and Elmira in a ravine. Also it grows at Roricks Glen, and is quite abundant in the shaded ravines of the northern portion of Bradford co. Pa., this seeming to be about its northern limit extending westward through Indiana to Illinois. This plant is most characteristic as to its habitat, only found in rocky, cool ravines, near water, extending southward according to the manuals, yet I am rather led to infer that its geographical range is limited through the middle states of the U. S., for it does not appear again in the Rocky Mountain flora. It should be found in Livingston and Cattaraugus counties, where the elevation is not too great.

The Creek Bottoms. This comprises the beds of the main creeks of the county with their lesser tributaries, where the species will vary according to soil and the increasing elevation of their course with such plants and schrubs, as *Salix lucida*, one of our handsomest willows, and *Lobelia cardinalis*. These creek beds are the home of the willows.

The woodlands. This comprises all the higher elevations of the county where any forest shade is left to protect the species peculiar to this region. *Salidago caesia,* many asters may be found here, *Lespedeza Stuvei* and *Habenaria Hookeri* have only been found on Sullivan Hill. with the trees and plants characteristic of the uplands.

Swamps and low grounds. Not all of these have been visited, one of the most interesting is Lowmans swamp in the township of Chemung, also Greatsingers swamp on Baldwin creek, the two most thoroughly explored. In the latter can be found *Geum rivale* which I also found in the swamp at the head of Christain Hollow, Bradford co. Pa., with *Abies balsamea.* The former is quite rich in its ferns, and those species peculiar to cool, well shaded bogs. *Habenaria psycodes, Mediola virginica, Quercus bicoler, Viburnum lantanoides,* etc.

The ponds. Here are found the usual species of the genera, *potamogeton, carex, juncus,* which the limits of this hasty sketch will not permit me to mention. Suffice it to say that I have now sufficient material on hand for a forth coming bulletin of the Chemung county flora, having up to the present time determined some 652 species, the grasses and sedges being yet to be more largely collected. This list has accumulated during the odd chances for gathering during the time from 1879 when the first main collections were made. This being my own unaided work is necessarily still yet very incomplete, and only those who have searched the hills or among the treacherous bogs can understand how much ground has to be covered to complete the work, every plant of my list having gone through my own hands.

This portion of the flora of New York seems most closely allied to that of the states of Ohio and Indiana and also Michigan, perhaps more uniformly so than other portions of the state, as that of the south-east corner of the state drifts into the coast flora of the eastern states and New Jersey. Michigan is listed with 1634 species, of which 627 can be found in Chemung, Indiana with 1432, with 564 for Chemung. From Lloyds exchange list of 484 of the most common species growing around Cincinnati, Ohio, 279 belong in Chemung. This showing a fairly good flora for so small a county, only yet partially worked over.

Pycnanthemum incanum

PROCEEDINGS

ADJOURNED ANNUAL MEETING

of September 22nd, 1891.

The Vice-President, Mr. I. B. Coleman, in the chair.

Minutes of the last meeting read by the Secretary, Mr. Robert A. Hall, and approved.

Report from the President, Mr. John R. Joslyn, was passed until he should have arrived.

Report from the Council of Administration, Prof. D. R. Ford, Chairman, was presented by him and accepted. The suggestion from the Council, that at the announcement of each meeting ten or fifteen minutes should be devoted to a presentation of any curious or new objects or specimens by any member, and have it examined and described, and named if needful, was adopted. The presentation, etc., to be called an " *Inspection Exercise.*"

The Treasurer, Mr. W. A. Eastabrook, presented his report for the year, showing a balance in the treasury of $620.00, which was accepted.

President Joslyn having arrived, took the chair, and read his annual report on the state of the society—briefly reviewing the successful results of the work of the past year, in which had been presented to the Academy and the public, subjects covering many branches of scientific study, all in the direction of " Discovery and Diffusion of Scientific Knowledge," which is the purpose and motto of the Academy. The report in full may be found in the published " Proceedings " of the society.

Under " Miscellaneous Business " the recommendation of the Council of Administration for the holding of meetings of two sorts—one month a *Popular meeting,* with a lecture or two illustrated papers or essays, by means of home or foreign talent, and again to hold a *Section meeting* on the next and alternate months, to bring out our *home* members, in a contribution of such fragments as they like in any science that interests them or has come under their notice, was adopted.

The election of officers for the ensuing year being next in order the Academy proceeded to ballot for the same. The following were declared elected :

PRESIDENT—MR. JOHN R. JOSLYN.
VICE-PRESIDENT—MR. ISAIAH B. COLEMAN.
SECRETARY—MR. ROBERT A. HALL.
TREASURER—MR. WILLIAM N. EASTABROOK.
TRUSTEE FOR FOUR YEARS—PROF. DARIUS R. FORD.

COUNCIL. { PROF. DARIUS R. FORD.
MR. ROSWELL R. MOSS.
MR. JABIN A. SECOR.

CHAIRMEN OF SECTIONS.

ASTRONOMY—PROF. DARIUS R. FORD.

MICROSCOPY—MR. RICHARD L. GUION.

GEOLOGY—MR. ROSWELL R. MOSS.

NAT. HISTORY AND BOTANY—MRS. C. F. CARRIER.

SOCIAL SCIENCE—MR. CHARLES M. MARVIN.

METEOROLOGY—MR. JABIN A. SECOR.

CHEMISTRY AND PHYSICS—MRS. J. R. JOSLYN.

ADDRESS of the President at the Annual Meeting, *June 9th, 1891.*

At the end of thirty years from its organization, the Academy of Science is again assembled in its annual meeting. It is stated in the certificate of incorporation, that the society was formed for scientific purposes, and that the particular business of the society is the pursuit of astronomical and scientific studies generally. In complying with the rule, that at the annual meeting the president shall make a report on the state of the society, it is plain that the first business is to note how faithfully the purpose of the founders of the society is carried out, next as to the interest of members in the work done, and only lastly as to the financial conditions of our Academy.

A public meeting has been held in our Observatory hall once in each month from and including October, 1890, to May, 1891, the period which constitutes the active term of our Academic year. Dr. T. F. Lucy opened the season in October, 1890, with a lecture on "History of Photography, Ancient and Modern." This was followed in November by a symposium on "Early Elmira," and so through the year we have been instructed by these gentlemen, and by Mr. W. F. Hopkinson on Comets and Meteors, by Mr. F. Collingwood, on Streets and Pavements, by Dr. VanNorden, on the New Psychology and Hypnotism, by Dr. Ford, on the Chronograph, and by the Rev. D. W. Smith, of Watkins, N. Y., on The Possibilities of Microscopical Photography. These lectures and papers are cited by title to show the kind of work done in the Academy during the year. Meanwhile the matter has been prepared for a pamphlet of about 72 pages devoted to setting forth some features of the Academy and its history, together with papers read in our hearing and descriptions of scientific instruments, invented and constructed by some of our members.

This pamphlet will be on the press in a short time and will be sent to all the leading scientific bodies in the world. So far as I can learn it is the first publication of our society, of any pretensions at least, and we shall certainly be very proud of it. Its publication will give us a standing and recognition among other scientific societies that we have never had, evidenced in the first place by receipt of their publications and the extension of their hands in fellowship, and instead of waiting many years before further publishing it is the hope of the present board of officers that those elected to conduct our affairs during the next year will begin the annual publication of our proceedings. The library of the society is growing slowly, thanks chiefly to the gentlemen who have represented this district in Congress. No doubt it will increase more rapidly after our publication introduces us into the first scientific circles—as it surely must.

As to the membership, it slowly increases. For many years no effort has been made to that end. Having money enough and to spare, according to our former usages, citizens were not urged to unite with the society. Still the work done and the opportunity here presented for scientific study and questioning, the free parliament that we have always maintained have drawn the attention of many thoughtful men and women to the value of the Academy and the wisdom of its maintenance. Never as I trust, will our society seek to win supporters and members by solicitation or advertisement. If there's honey in this hive those who love that kind of honey will find it out, and seek it, and if there's no honey the hive will be neglected as it ought to be. I like the pride and dignity that will cordially open its doors and invite all to share its treasure, but will solicit none, at a dollar a head.

But more care will be taken than heretofore to keep a list of members and to include that list in our publications, and to meet the requests of those who are so interested as to seek a membership with the society. We are proud but not indifferent.

As to our financial condition it is ideally sound. "Owe no man anything" says the Apostle, "save only for the *love* you ought to owe." That's precisely our condition. We have no debt, of any description. We have put into the hands of the Elmira College real estate and scientific instruments of the value of about $10,000.00, reserving to the society the perpetual use of

both, and we have $600.00 in the hands of the Treasurer. We
can use money wisely, but we won't pass the hat. It is my
hope that during the year the fellows of the society will claim
the privilege of contributing the same fee as the annual member,
that is $1.00, so as to keep the present funds of the society
intact, but enable us also to engage distinguished men residing
in our vicinity to come here and talk to us about their favorite
studies. For there's this beautiful feature in our scientific
friends, they will come here if we will but pay their traveling expenses. You can see at once what lessons of interest are at our
command for trifling cost.

Now but a word as to the plans for the ensuing year, for I
don't want to speak merely of the dead past as showing the state
of the society. We have more to do with what is ahead of us.

We have confined our public meetings of late years, very
largely to lectures and essays. I suspect that for the next season's work the Council of Administration will recommend an
alternation of lecturing with studying. A lecture one month,
the next month to bring on work in charge of one of the sections.
A year ago some of us were a little alarmed over the outlook, to-
day we perceive opportunities on every hand, and though in our
membership we lack for those given professionally to scientific
studies, if we but continue our work faithfully the Lord, in the
favorite dining room motto, the "Lord will Provide." On the
whole then, the society is very flourishing. A young officer
rushed up to Wellington once with the glad news "my Lord, we
have taken a standard." "Go back and take another was the re-
ply." Let us continue in well doing and be grateful for the
equipment that we have however much we better it.

J. R. JOSLYN,
President.

TITLES OF ORIGINAL PAPERS read before the Elmira Academy of Sciences.

"The cause of the Recession of the Waters of the Gulf of Mexico."—By Rev. David Murdoch, D. D.

"History and Progress of Natural Science."—By Prof. C. S. Farrar.

"Vegetable Respiration."—By Dr Wm. M. Gregg.

"Ethnology."—By Prof. J. E. Latimer.

"How does Nature sow Her Forests?"—By Prof. I. M. Wellington.

"Progressive Development or Transmutation of Species."—By Prof. J. E. Latimer.

"Antiquity of Human Remains."—Rev. A. W. Cowles, D. D.

"Has the Earth ever Revolved about a different Axis from its present one?"—By Mr. Francis Collingwood, C. E.

"Geological explanation of the forming of the Crust of the Earth."—By Prof. C. S. Farrar.

"Longevity."—By Mr. Francis Collingwood, C. E.

"Asteroids and the searches for them."—By Prof. C. S. Farrar.

"Geology of Canada"—By Mr. Augustus McConnell.

"Do Insects feel Pain?"—By Dr. Wm. H. Gregg.

"History of Climates as investigated and recorded by the Ancients, beginning with Ptolemy."—By Prof. I. M. Wellington.

"The Sun, with particular reference to the spots that appear and disappear on its disc."—By Prof. C. S. Farrar.

"Musical Chords."—By Mr. Francis Collingwood, C. E.

"Ivory, how and where obtained, also curiosities of its manufacture."—By Mr. Robert Hall.

"Economics of the Gulf Stream."—By Rev. Wm. Bement.

"Gunnery."—By Rev. Thos. K. Beecher.

"Small Pox, Vaccination and its prevention of the Disease."—By Dr. Ira. F. Hart.

"Asteroid, particularly the Asteroid "Echo.""—By Prof. Rogers, of Alfred Academy.

"Coal Oil, its Chemical Properties, &c."—By Prof. C. S. Farrar.

"The Brown Rat."—By Prof. I. M. Wellington.

"Tycho Brahe."—By Prof. C. S. Farrar.

"Trees, electrically considered."—By Prof. I. M. Wellington.

"Heresies of Darwin."—By Prof. J. E. Latimer.

"The Planet Stars."—By Mr. A. McConnell.

"The Law of Rest and Motion."—By Mr. Francis Collingwood, C. E.

"Habits and peculiarities of the Crow Black Bird (Molothus Peccoris.")
—By Dr. Wm. H. Gregg.

"Papyrus and Paper."—By Mr. R. A. Hall.

"Steam Valves and Cutoffs."—"Cotton and the Cotton Gin."—By Prof.
J. E. Latimer.

"Geometry of Plants."—"Inter-changeableness of Physical Force"—By
Prof. D. R. Ford.

"The necessity of Forest Culture in the United States."—By Mr.
Francis Collingwood, C. E

"Anaesthetics."—By Prof. D. R. Ford.

"The Philosophy of Colors."—By Prof. D. R. Ford.

"Modern Applications of Electricity and its use as a Motor on Railways."—Illustrated by a small working model.—By Prof. D. R. Ford.

"Woodchucks."—By Dr. S. O. Gleason.

"Horology."—By Rev. T. K. Beecher.

"History of errors in Microscopic work."—By Dr. Adele A. Gleason.

"The Microscope in legal evidence."—By Dr. H. D. Wey

"Alaska. its climate, scenery and peculiar people."—By Mr. Francis
Hall.

"Resolving of Crude petroleum into its 8 or 9 parts."—By Prof. D. R.
Ford.

"The method of calculating an eclipse of the Sun."—By Miss Alice
Hall.

"Free Trade."—By Prof. Charles A. Collin.

"One Phase of the Tariff question."—By Mr. Jay S. Butler.

"Flora of Chemung County."—By Dr. T. F. Lucy.

"How children should study Nature."—By Mrs. Mary P. Joslyn.

"Electric Welding."—By Mr. I. B. Coleman.

"Adulterations in articles of food."—By Mr. Clay. W. Holmes.

" Mind cure, and Kindred Delusions."—By MR. FREDERIC HALL.

" Weregild, or the Pecuniary value of the Social Atom."—By MR. R. R. MOSS.

" Role of Micro-organism in disease."—By DR. F. W. ROSS.

" The great freshet in the Chemung River June 1st, 1889."—By MR. JABIN A. SECOR.

" The Water Shed of the Tioga and the Chemung rivers, and the recent great freshet in them."—By MR. R. A. HALL.

" Africa and Livingstone's Travels."—By PROF. P. W. LYON.

" Glaciers and the Glacial Period."—By PROF. H. L. FAIRCHILD.

" Bicycles and the trip of the American Bicycle Club to Europe."—By MR. W. N. EASTABROOK.

" The Edison Phonograph."—By MR. FRANK E. BUNDY.

" History of Photography, both Ancient and modern."—By DR. T. F. LUCY.

" Street Pavements."—By MR. FRANCIS COLLINGWOOD, C. E.

" Comets and Meteors."—By MR. W. F. HOPKINSON.

" The new Psychology, and Hypnotism."—By REV. CHARLES VAN ORDEN, D. D.

" Possibilities of Microscopical Photography."—By REV. DWIGHT W. SMITH.

" Geology of Chemung County."—By MR. A. M. McCONNELL.

www.ingramcontent.com/pod-product-compliance
Lightning Source LLC
Chambersburg PA
CBHW032016190326
41520CB00007B/496